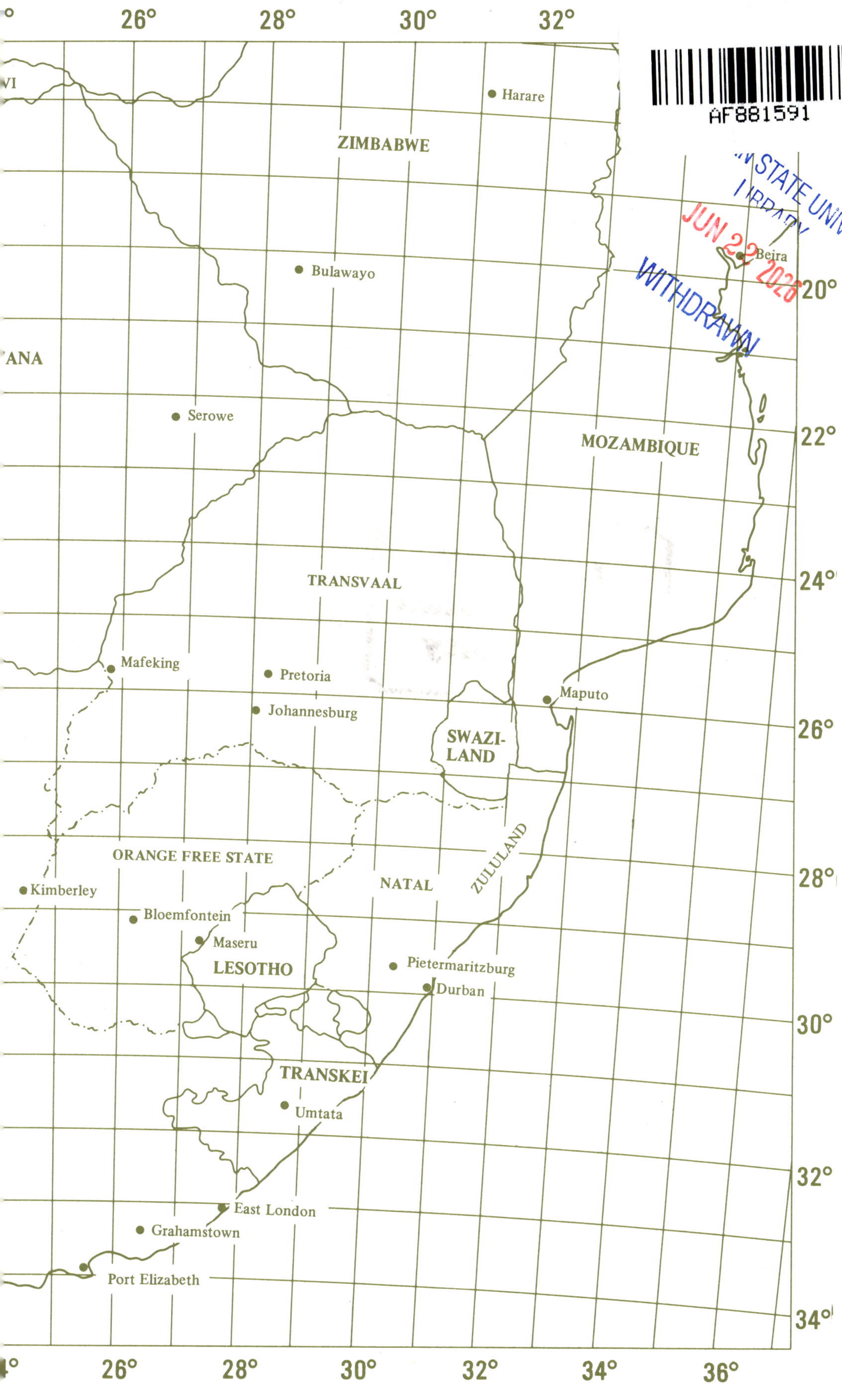
26°
28°
30°
32°
Harare
ZIMBABWE
Bulawayo
Beira
20°
ANA
Serowe
22°
MOZAMBIQUE
TRANSVAAL
24°
Mafeking
Pretoria
Johannesburg
Maputo
SWAZI-
LAND
26°
ZULULAND
ORANGE FREE STATE
NATAL
28°
Kimberley
Bloemfontein
Maseru
LESOTHO
Pietermaritzburg
Durban
30°
TRANSKEI
Umtata
32°
East London
Grahamstown
Port Elizabeth
34°
26°
28°
30°
32°
34°
36°

CEROPEGIA, BRACHYSTELMA AND RIOCREUXIA
IN SOUTHERN AFRICA

Plate 1. *Brachystelma barberae* (species 59); plant in Queen Elizabeth Park, Pietermaritzburg, Natal: 'the plant was simply an umbel of flowers sitting on the ground; the flowers look delightful but smell vile', *O.Hilliard* (photo: J.Stewart)

CEROPEGIA, BRACHYSTELMA AND RIOCREUXIA IN SOUTHERN AFRICA

R.ALLEN DYER
Formerly of Botanical Research Institute, Pretoria

With illustrations by
AURIOL BATTEN, ROSEMARY HOLCROFT & OTHERS

A.A.BALKEMA/ROTTERDAM/1983

ISBN 90 6191 227 X

Published by A.A.Balkema, P.O.Box 1675, 3000 BR Rotterdam, Netherlands

Distributed in USA & Canada by: MBS, 99 Main Street, Salem, NH 03079, USA

Printed in the Netherlands

86-6576

Foreword

Southern Africa is a treasure-house of plants of such variety of form and colour that its equal is hard to find elsewhere. This abundance has brought our flora international fame but is at the same time one of the reasons why it is still not sufficiently known. More than 20 000 species of flowering plants are found in southern Africa, the area south of the Kunene, Okavango and Limpopo rivers. Even though intensive research is being undertaken by a large number of local and overseas botanists, the Flora of Southern Africa, which is to replace the useful but outdated first general inventory, the 'Flora Capensis', will not be completed before the end of the century.

Expanding the knowledge of our flora is a time-consuming and intricate task to which the author of this book has already made a major contribution. In the classic volumes on the succulent Euphorbieae and many smaller papers on the Stapelieae, Allen Dyer clearly demonstrated his interest in the succulent plants in which southern Africa abounds. His studies of *Brachystelma, Ceropegia* and *Riocreuxia* represent an extension of this early interest. In these genera the succulent habit, expressed in the tubers, roots, and stems or leaves of many of their members, is accompanied by some of the most intricate and bizarre floral structures to be found in our flora. As pointed out by the author, the relationships between the different species and their insect pollinators is a fascinating story yet to be unravelled.

This book, the popular version of the accounts of these genera recently published by the author in the Flora of Southern Africa, represents the culmination of many decades of interest in the succulent members of the family Asclepiadaceae. Even so, due to the extreme rarity of some of the species of these three genera and the sparsity of many of the populations, quite a few of the taxa are still not sufficiently known. Thanks to the most persistent and painstaking work done by an experienced and dedicated botanist, assisted by many field workers, these genera are now becoming better known. We are much indebted to Allen Dyer, who even after his 'second' retirement at the age of 79 years, continued his work and completed this book, which is sure to become a standard reference work for all those interested in succulents and exotic floral structures.

Like the author, I hope that this book with its wealth of information will promote the knowledge which leads to deeper enjoyment and the urge to conserve rather than to destroy by the mania of acquisitiveness.

B.de Winter
Botanical Research Institute, Pretoria
January 1983

Contents

List of plates

Introduction

It is not long since I completed an account of the genera *Brachystelma, Ceropegia* and *Riocreuxia,* in the family Asclepiadaceae, for the Flora of Southern Africa, volume 27,pt.4 (1980). The use of the word completed could be misleading. Research is complete only up to a point and in the case of these genera new information has been added since the publication of the Flora and incorporated in the present account. The present text is not the result of specialization on the genera but merely the data accumulated at odd times during a long career in the Botanical Research Institute. It is possible for me to write about only those species occurring in the region covered by the Flora of Southern Africa, namely South West Africa (Namibia), Botswana, Lesotho, Swaziland, the Republic of South Africa, Transkei and the other emergent Black states. Many attractive species occur beyond these confines in Tropical Africa and further afield in the Orient.

My object now is to present the available information in a less formal, more pictorial manner than in the Flora. The emphasis is on illustrations to facilitate the recognition and appreciation of the individual species by both botanists and amateurs. An effort has been made to strike a balance with the text of the Flora and to avoid unnecessary duplication. The keys to the species have been retained very much as they were, based unavoidably in the final stages, largely on the vital internal structure of the flowers. At the same time these intricate details have been omitted from the formal descriptions of individual species. Where appropriate, the notes on individual species have been expanded for greater interest in the environment. To compensate the keener readers for some of the shortfalls in the text the illustrations contain fuller information.

Some readers would wish to know the derivation of the generic names: *Brachystelma* is not altogether clear. It comes from Greek meaning short crown or girdle and could refer to the corona or the girdle-like base of the corolla as in *B. tuberosum,* the type species of the genus from the Cape. *Ceropegia* comes from Greek meaning wax and fountain, which is the impression given by the many-flowered, regular looking inflorescences of *C. candelabrum,* the type species of the genus from Sri Lanka (Ceylon) and India. *Riocreuxia* is named after Alfred Riocreux, a skilful French botanical artist living from 1820-1912.

The interrelationship of the three genera is close in the tree of evolution, from *Riocreuxia* to *Ceropegia* and from *Ceropegia* to *Brachystelma.* The treatment here of the three genera in the reverse order as in common usage from *Brachystelma* to *Ceropegia* to *Riocreuxia* is of no significance and *Ceropegia,* the centre of interest, is in the middle either way.

To the succulent plant enthusiast the Ceropegias need little introduction. Some species have evolved a completely succulent habit and others are completely herbaceous. Brachystelmas and Riocreuxias are less widely known. Brachystelmas are perennial herbs with no succulent representatives but most have a tuberous root and the flowers are often reminiscent of representatives of the Stapelieae. This likeness caused the eminent Swedish botanist, Carl Thunberg, who travelled extensively in the Cape, 1771–1773, to describe the first species of *Brachystelma* he collected as *Stapelia caudata (= Brachystelma caudatum,* species no.3).

In *Riocreuxia,* which has at times been included in *Ceropegia,* the species are mainly vigorous herbaceous climbers with large heart-shaped leaves and relatively long dense sprays of delicate tubular flowers. They prefer to grow unrestrained, which does not suit the cultivator who has only limited space.

For those collectors and growers who demand a specific name for each and every plant, wild and cultivated, this account is bound to be a disappointment. A few species are distinctive and identifiable at a glance; others show variations from habitat to habitat and even within a single population. Mixed populations of related species in one habitat may give rise to the most complex problems in taxonomic classification, one factor possibly being natural hybridization and another, natural variability. Harry Hall records an amazing result of a changed environment in the case of *Ceropegia occidentalis* from the isolated habitat of Papendorp near Vredendal on the west Cape coast. 'Three years ago (1976) I removed a small lateral tuber to give it horticultural treatment, no flowers having been seen on the upright plant in the wild for 8–9 years. At Hermanus it grew apace, quickly coiled its stems around supports and produced flowers identical with the early ones from Papendorp. It has flowered for over 2 years without a pause having an average of about 12 flowers daily.' *Ceropegia woodii* known and grown throughout the world as an ideal subject for a hanging basket and whose branches are seen to dangle and flower freely, can do otherwise. A specimen collected not far from the type locality near Durban, Natal, and appearing in every way typical, when first introduced to cultivation at BRI, responded by producing a pendulous branch where no support existed and a spiral climbing branch on a support provided on the other side. Many similar instances of natural variability could be cited.

All plants are interesting, but for some people the Ceropegias and Brachystelmas will always have a special fascination. It is no real compliment that my article on *Encephalartos* in Bothalia 8, 4:405–515 (1965) caused a veritable acquisitive explosion in the Cycads of Southern Africa. The trade has continued almost unabated but under greater control by nature conservation bodies. Will the same rate of possessive interest be engendered in the public on the present three genera by this publication? I hope not. But disciplined interest between the collector and the botanist does result in the recording of a wealth of useful scientific data and this is to be encouraged especially for poorly documented species of which there are still very many examples.

ACKNOWLEDGEMENTS AND THANKS

In preparing the account of *Brachystelma, Ceropegia* and *Riocreuxia* for the Flora of Southern Africa, 27:4 (1980) I received an abundance of help. The Directors and Curators of overseas and local botanical institutions sent type specimens and study material on loan. I owe them all my sincere thanks. I am most grateful to David Field at the Royal Botanic Gardens, Kew, who has given me valuable assistance with the identification of specimens of *Ceropegia* with tropical African affinities and of an undescribed species from Swaziland. I also wish to thank all my colleagues, collectors and other friends who have assisted me in one way or another over the years to this final stage of preparation. Among these there is no first and foremost no last but not least. One cannot evaluate each contribution – they are all important links in a long chain. I must, however, mention a few by name and apologise to those whose names were squeezed out in the cause of brevity.

I am most grateful to Dr B.de Winter, Director of BRI, for very generously allowing me to complete this work with all the facilities and accommodation of the Botanical Research Institute, before and after my second retirement at the end of September 1979, and for kindly writing the Foreword.

Auriol Batten, joint author and artist of the book on Wild Flowers of the Eastern Cape Province, 1966 and other illustrated works, lends colour and distinction to the present work by her several contributions of beautifully executed paintings. Rosemary Holcroft, official artist on the staff of BRI, has contributed several original drawings and has brought back to life artists of bygone days by adapting old plates to give more uniform coverage of species than would otherwise have been possible.

The work of official photographers, past and present of the BRI from Harry King and Jimmie Reyburn to the present occupant, Adela Romanowski, has been invaluable in the compilation of this account.

Roy and Estelle Bayliss have made several contributions of species new to science and have rediscovered others thought to be on the verge of extinction. Harry Hall, formerly in charge of the succulent collection at Kirstenbosch, has also made several important discoveries which are enlarged upon in the appropriate places. Bruce Bayer, present curator of the Karoo Botanical Garden, at Worcester, Cape, has generously shared his field knowledge, notes and material to the benefit of this work. Dave Hardy who has been in charge of the living collection of succulents of the BRI for many years and Carole de Wet and her assistants, have successfully liaisoned between the photographer and myself.

Olive Hilliard and Bill Burtt in their search for typical *Brachystelma pulchellum* made two new discoveries in *Brachystelma ngoyanum* and *B.petraeum* in Natal and Percy Lancaster has added *Brachystelma gemmeum* in the Transvaal.

The early interest of my colleague Les Codd, in *Ceropegia* particularly, is evident in the large number of his meticulously kept records preserved in the National Herbarium, BRI. E.A.Phillips, horticulturist, is another who contributed extensively to BRI in the past. If one digs deeper the names of H.R. Brownlee, Florence Paterson, A.G.McLoughlin, F.G.Carnegie, T.N.Leslie, Eileen Bruce and others reappear from the distant past while Peter Bruyns is an up and coming enthusiast.

I must call a halt now and express my sincere thanks to Dr Otto Leistner, who edited my work on these three genera, for the Flora of Southern Africa, vol.27, pt.4 (1980), who dotted the i's and crossed the t's and in so doing rendered a valuable service indirectly to the present account.

NOMENCLATURAL MATTERS

The nomenclature used here conforms to the International Botanical Code. In classifying the multitude of specimens which come to hand one would like to be or even appear to be consistent. But there are many pitfalls because of the variability of nature and such problems will be discussed in appropriate places. A few peculiarities in nomenclatural practice seem to warrant a short comment here.

Since the generic names *Ceropegia, Brachystelma* and *Riocreuxia* all end in 'a', it is rather natural to expect that the specific names or epithets would be consistent in their endings, that is, if the specific ending is 'a' in *Ceropegia* the other genera would be the same, but it is not so. It is *Ceropegia africana;* it would be *Brachystelma africanum* and *Riocreuxia africana. Brachystelma,* the odd one out, is neuter gender in Latin (not feminine as the other two) and that causes the apparent, but not real, inconsistency. One just has to get used to these exceptions.

In a few instances it has been found desirable to recognise subspecies: some might prefer varieties. The point to be made is that the term variety is used in the horticultural trade rather indiscriminately, which has discredited its use in taxonomic botany. Thus where applicable I have changed the status of varieties in Huber's work on *Ceropegia* to subspecies. In the case of the specific name *Ceropegia stapeliiformis* if one wishes to distinguish a particular variant as a subspecies, then the original concept automatically takes subspecific rank by the repetition of the specific name, thus the first subspecies is automatically *Ceropegia stapeliiformis* R.Br. subsp. *stapeliiformis,* and one chooses any others for example *C.stapeliiformis* R.Br. subsp. *serpentina* (Bruce) R.A.Dyer.

The name or abbreviation following the botanical name denotes the person or author responsible for the botanical name. Some authors are given their surnames, which may be abbreviated if it is lengthy; others are given their initials. For instance it is N.E.Br. and not Brown, while Schlechter is abbreviated to Schltr. N.E.Br. is not to be confused with R.Br., Robert Brown, while Schlechter has no opposition in the botanical field and requires no initials. Linnaeus, the father of our botanical nomenclature is simply L.

To avoid confusion with Sir William Turner Thiselton-Dyer, KCMG, etc., former Director of the Royal Botanic Gardens, Kew, editor of the Flora Capensis 1896–1925, after whom *Aloe dyeri* and other species were named and with whom I have no blood relationship, I am R.A.Dyer in botany and sometimes sign R.Allen Dyer, because of the many other unrelated persons of the same surname.

FERTILIZATION, SEED DISPERSAL AND CULTIVATION

In the family Asclepiadaceae it is not the honey bee which does the fetching and carrying of the pollen masses from the stamens to the stigma but various species of fly and other insects. The family has evolved a highly specialized mechanism for the fertilization of its flowers, in fact it is mainly on the unique mechanism for pollination that the family is circumscribed.

An informative article on the 'structure of the highly specialized carrion-flowers of Stapeliads' by G.Reese appeared in Cactus and Succulent Journal (US) Vol.45:18 (1973). The information on fertilization applies equally well to the present genera which as mentioned are the forerunners of the Stapelieae.

Without going into the intricate details available in literature it may be stated that male and female reproductive parts (stamens and style) are united into a small columnar structure within the base of the corolla (the staminal column): the pollen is aggregated into waxy masses (pollinia) one mass or pollinium in each anther cell. Two pollen masses, one each from neighbouring anthers, are attrached by fairly tough, to very delicate, connectives to a common small hard carrier. The primary subdivision of genera is based on the position of the pollen masses in or on the staminal column, whether pendulous or suberect or erect. In *Brachystelma, Ceropegia* and *Riocreuxia* the pollen masses are suberect to erect, near or on top of the staminal column.

Insects are attracted from a distance by the odour of the nectar, sometimes sweet but usually putrid smelling. Once in the vicinity of the flower there are further attractants such as colour and hairs, which may be fixed or vibratile in a light breeze. A visiting insect, such as a small fly, may get its proboscis or a leg caught in the guide to the pollen masses which may then be extracted from the staminal column by a greater or lesser degree of force. Subsequently the insect may – rather problematically – deposit one or both pollen masses on one of the five stigmatic surfaces fused within the top of the staminal column, and by this means effect pollination.

Many collectors have observed small flies trapped in the inflated base of Ceropegia corollas by stiff deflexed hairs higher in the tube. This ensures that the flies have ample time to extract or to deposit pollen masses before obtaining their release when the flowers begin to wilt and the hairs lose their turgidity.

One automatically expects the evolution of the intricate mechanism of the staminal column and other refinements to go hand in hand with efficiency.

Figure 1. *Ceropegia conrathii,* x 1: to illustrate seed dispersal: the ripe pods or follicles, which usually develop in pairs, have split lengthwise and the numerous seeds are being released piecemeal, each with its expanded parachute-like tuft of hairs ready to be carried away by the gentlest breeze.

The results are quite the reverse. The percentage of successful fertilizations varies from genus to genus and from species to species but with few exceptions it is particularly low when compared with the bee-pollinated families of flowering plants. The weak link appears to be the actual placement of a pollen-mass on one of the somewhat hidden stigmatic surfaces.

Seed dispersal

Once fertilization has been successful, however, the efficiency pattern changes dramatically. There is a variable period of after-ripening during which the pedicel or stalk, which feeds the ovary, strengthens, sometimes from a threadlike structure into a strong upright support for the follicles or seed pods (Figure 60). Each follicle develops rapidly to full length but takes several to many weeks to mature the closely packed flattish seeds, each with a terminal tuft of hairs. On ripening the follicles split lengthwise releasing the seeds with their expanded parachute-like tuft of hairs which carry them near and far on the gentlest breeze (Fig. 1).

Figure 2. *Brachystelma barberae* with a tuberous rootstock and corolla lobes united at the tips; a. tuber with one annual stem and inflorescence, x ½; b. corona basin-shaped with 5 deep pockets, outer margin narrowly and deeply V-shaped, inner lobes arise from within the basin and are incumbent on the anthers, x 5; c. young follicles, x ½ (adapted from photograph by H.King)

Cultivation

Basically, a light sandy soil with a little old compost is likely to give satisfactory results. Ensure good drainage. Do not be lavish with artificial fertilizers, be frugal with watering generally, and pay heed to the natural resting periods. Each successful grower is likely to have discovered individual refinements and it will be wise to follow his or her advice until you have acquired your own know-how.

SURVIVAL FOODS

By comparison with their numbers few species of Asclepiadaceae have been incriminated as plant poisons. To the best of my knowledge no species in the three genera dealt with here has been suspected of poisonous properties, but I am unaware whether or not Riocreuxias are normally grazed by animals but there are frequent references to the leaves being used as spinach.

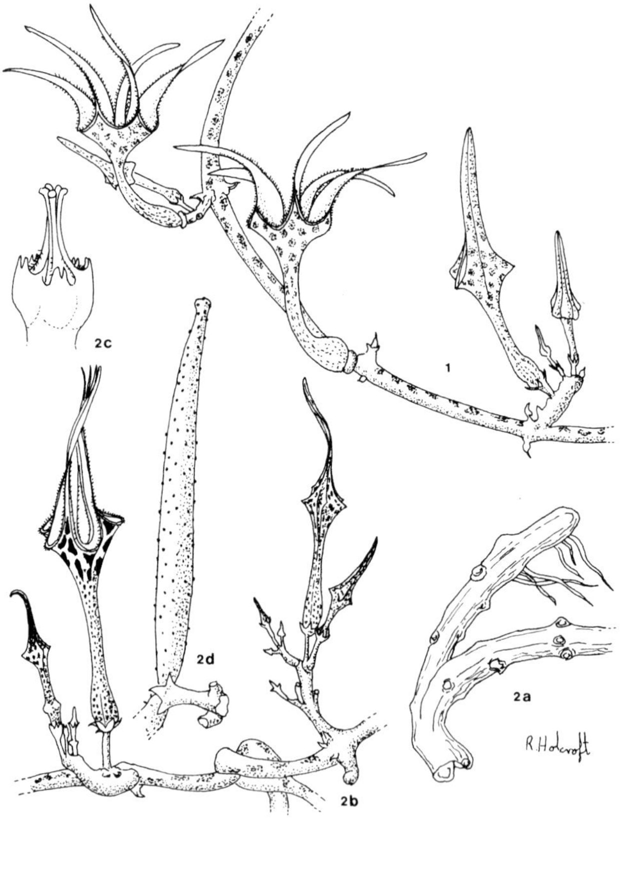

Figure 3. 1. *Ceropegia stapeliiformis* subsp. *stapeliiformis* with fibrous roots, wholly succulent stem and corolla lobes free at the tips, x ½. 2. subsp. *serpentina:* 2a. rooting stem, x ½; 2b. flowering stem with buds, x ½; 2c. corona with upright inner lobes, x 4; 2d. showing one of the two follicles or seed pods, x ½ (after Flower. Pl. Afr. t.809, 1072) (Flora S.Afr. 27,4:Fig.10 (1980)).

What is also known is that from ancient times some species in the family have been used for medicinal purposes. From time immemorial the tubers of Brachystelmas and Ceropegias have featured in the daily diet of the aborigines. Those Bushmen whose habits have not been greatly influenced by European contact still rely heavily for their survival on indigenous plants with edible fruits and tubers. Among Bantu it is largely the herd boys who, in their close contact with nature, keep up the habit of eating the tubers of *Brachystelma* and *Ceropegia.* There can be no doubt that the edibility of the tubers has influenced the distributions of the more sought after species, some of which are extremely rare. The prince of collectors, Ernest Galpin, and others have often referred to this situation. Had Roy Bayliss not observed Lesotho herd boys making a meal of tubers he might not have rediscovered the rare *Brachystelma perditum* (Plate 4, species no.5).

One must not forget either that wild animals such as porcupines, baboons,

Figure 4. *Ceropegia africana* with a tuberous main root, semisucculent leaves and climbing stem, corolla lobes united and distinctive corona lobes; a. the type of the species and of the illustration published in 1822, x ½; b. corona with falcate inner lobes (after Bot. Reg. 8: t.626, from the Cape but no exact locality)

rodents and insects are also fond of the tubers and possibly rival man in the extent of their depredations of the tuber population.

In my youth in Natal I have often eaten the succulent coronas with sweet nectar of dwarf species of *Xysmalobium* and *Schizoglossum* (both Asclepiadaceae) in grassveld where they would be grazed by animals in the normal course of events. I regret that I made no record of the actual species.

Thus the family is important from the aspect of survival foods. For anyone cut off from their normal food and water supply, as might well happen to troops engaged in bush warfare, the presence of *Brachystelma* and *Ceropegia* tubers might well make the difference between life and death.

KEYS TO GENERA AND SPECIES

The botanical keys which follow are verbal guides designed to assist those readers who have some prior knowledge of the groups concerned. The foolproof key is a botanist's dream, so the reader who does not meet with immediate success in using the keys has no cause to be disheartened. If the reader is unfamiliar with such keys he should turn first to the illustrations and text and there obtain quicker results and greater enlightenment.

The arrangement of species in the keys indicates to a greater or lesser degree the relationship of species to species but for the final analysis one must consult the text.

A feature of the keys to the species of *Brachystelma* and *Ceropegia* is the primary importance placed on the root-system. This should appeal to the grower for he knows that a plant, which normally produces a tuber when grown from seed, always produces a tuber. It does not produce a tuber one time and a cluster of succulent fibrous roots at another. The keys include little or no reference to leaves because of their general variability, even from heart-shaped to linear in one species or even on one plant. In the main keys to species are based on floral characters which are relatively stable in structure.

Pollinia, all important in the process of fertilization, are on the average approximately eight times smaller than the head of an average pin. It seems probable that an accurate knowledge of the pollinia would be of value in classification of the species, and consequently in the keys, but so far it has not been found practical to examine and illustrate all species in their three dimensional complexity. Released from the tensions within the anther cells the pollinia usually take up altered positions and are difficult to manipulate. There are differences in shape of the outline, colour and marking of the pollinia, in

---

Plate 2. *Riocreuxia torulosa* (species 2); young flowering branch with many-flowered inflorescence, x 1 (*Auriol Batten,* Mount Kemp, eastern Cape)

Auriol Batten

their thickness and in the length and curve of the pellucid inner or upper margin. The unknowns are legion.

KEY TO GENERA *BRACHYSTELMA*, *CEROPEGIA* AND *RIOCREUXIA*

Flowers solitary to several, rarely many, in clusters or umbel-like cymes; corolla wide open to tubular; anthers incumbent, rarely suberect; pollen-masses mainly pellucid along upper part of inner margin, with carrier often winged:

- Corolla opening nearly flat or tube campanulate; when tubular, then less than twice as long as its diameter at middle (see *Brachystelma oianthum*); plants mainly dwarf with tubers or a few with a tuft of fleshy roots . . . . *Brachystelma,* p.12
- Corolla tubular with the tube usually 2 to several times as long as its diameter at middle, usually with inflated base; plants mainly climbers or scramblers, sometimes dwarf, with tubers or tuft of fleshy or fibrous roots . . . . *Ceropegia,* p.133

Flowers in open branching cymes or panicles, usually with few to many clusters scattered along its branches or racemose along a single peduncle, rarely sessile; corolla tubular in lower half, usually slightly inflated or bell-shaped; anthers ± erect; pollen-masses pellucid at apex; roots fleshy or fibrous; stems herbaceous, usually climbing; leaves usually large and heart-shaped . . . . *Riocreuxia,* p.227

Brachystelma

Brachystelma *R.Br.* in Curtis's Botanical Magazine t.2343 (1822); R.A.Dyer in Flora S.Afr. 27,4:1 (1980) [Fig.5]

The following names which are met with in literature are regarded as synonyms of *Brachystelma: Decaceras* (1863); *Dichaelia* (1868); *Micraster* (1868); *Lasiostelma* (1876); *Tapeinostelma* (1893); *Craterostemma* (1893); *Brachystelmaria* (1895); *Aulostephanus* (1896); *Blepharanthera* (1913); *Siphonostelma* (1913).

The large number of these synonyms is an indication of the difficulties in classification and the uncertainty which has prevailed in the minds of different workers on the question of generic limits. Several of these generic names could be resuscitated if one were so minded and could well find a following. For me it would have no real advantage since the interrelationships are diverse, complicated and beyond my comprehension.

The genus *Brachystelma* was founded by Robert Brown in 1822 under the name *Brachystelma tuberosum* but even then the species was recorded vaguely as from the Cape of Good Hope. Further information will be found under the species heading no.9.

Description: Perennial herbs, usually dwarf, with a single tuber and sometimes with auxilliary tubers and rootlets or a cluster of fleshy roots, producing 1 to several annual stems from the base. *Stems* prostrate to erect, branched or unbranched. *Leaves* opposite, with or without stalks. *Flowers* 1 to several together at the side of stem-joints or rarely terminal, with short to medium lengthed stalks. *Corolla* divided to half way or further to base (but see *B.oianthum*); lobes 5, free at the tips and variously spreading, or united at tips forming a cage-like structure.

Species connecting the genera *Brachystelma* and *Ceropegia* include *Brachystelma oianthum* (species no.2) and *Ceropegia mafekingensis* (species no.28) and further information about them will be found under their respective headings.

Brachystelma species occur from Africa to India and Burma and further East, but predominantly in southern Africa. The present total is about 100 but it is surmized that the number would be appreciably increased by intensive field work over the full range of distribution. The number of previously undes-

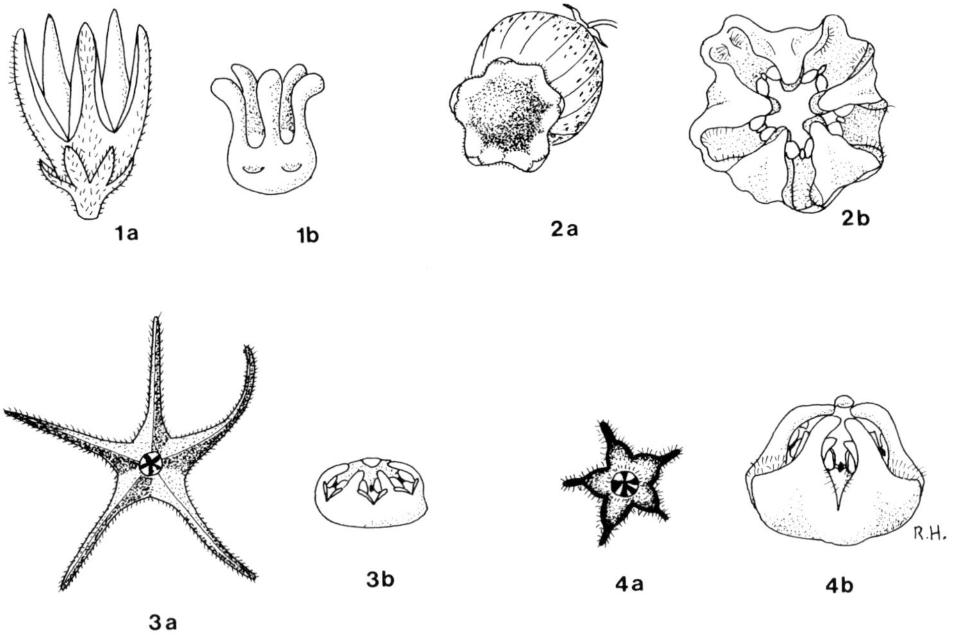

Figure 5. 1. *Brachystelma blepharanthera:* 1a. flower, x 3; 1b. corona, 1-seriate, shortly cup-shaped at base, the equivalent of the outer series is the thickened outer margin of the 5 pockets, confluent with the base of the inner series, the inner lobes erect, slightly recurved in upper half and considerably overtopping the staminal column, x 10. 2. *Brachystelma oianthum:* 2a. flower, x 1; 2b. corona with shallowly basin-shaped base, with outer margin of the 5 pockets confluent with the base of the inner series, inner lobes broad-based, incumbent on the backs of the anthers (not meeting over centre of staminal column), x 10. 3. *Brachystelma caudatum:* 3a. flower, x 1; 3b. corona with campanulate base with outer margin of the 5 pockets V-shaped, confluent with the base of the inner series, inner lobes incumbent on backs of anthers, x 5. 4. *Brachystelma tuberosum:* 4a. flower, x 1; 4b. corona appearing uniseriate, with campanulate base with outer margin of the 5 pockets narrowly V-shaped, confluent with the base of the inner series, inner lobes incumbent on backs of anthers, x 10 (adapted from BRI records)

cribed species in southern Africa recorded in the past few years is about 20 and there must be more awaiting discovery for the first time in science. At the same time there are a number of species which have been recorded only once in 200 years of exploration.

KEY TO THREE MAIN GROUPS OF SPECIES OF *BRACHYSTELMA*

1 Rootstock producing a tuber:
  2 Tips of corolla-lobes free . . . . – **Group A** (1-46)
  2 Tips of corolla-lobes united, at least when flower first opens (sometimes free or severing in *B.pygmaeum, B.dimorphum* and *B.pachypodium*) . – **Group B** (47-59)
1 Rootstock producing a cluster of fleshy cylindric or fusiform roots . . . . – **Group C** (60-69)

Group A. Rootstock producing a tuber; corolla-lobes free at tips:

1 Corona developing from base or slightly above base of staminal column, with various modifications:

2 Corona appearing to be one-seriate, forming 5 nectariferous pockets opposite the pollinia and laterally confluent with the inner corona lobes, with the outer margin of pockets entire, emarginate, U- or V-shaped or with a narrow vertical slit, rarely extending into two short lobes:

3 Inner corona-lobes incumbent on the staminal column and not elongated above it:

4 Corolla-tube ovoid-campanulate, 10-15 mm long, contracted slightly to the mouth, with short lobes, long white hairs within; dwarf upright herb . . . . 2. **B.oianthum**

4 Corolla-tube campanulate or cylindric to absent, with lobes variably spreading:

5 Corolla divided to about two-thirds or more of its length:

6 Corolla 26-45 mm long to tips of lobes; flowers usually 2-several from a node, opening together:

7 Inner surface of corolla-lobes variably hairy, sometimes only minutely puberulous:

8 Inner corona-lobes simple, usually more or less lanceolate:

9 Inner surface of corolla-lobes variably hirsute, leopard-spotted and striped round corona . . . . . 3. **B.caudatum**

9 Inner surface of corolla-lobes puberulous, yellowish . . . . 4. **B.glenense**

8 Inner corona-lobes subquadrate, truncate, with dorsal ridge, meeting round centre of staminal column; upper surface of corolla-lobes minutely puberulous . . . . 5. **B.perditum**

7 Inner surface of corolla glabrous, dark purple or maroon, usually striped or spotted round corona; stems and leaves often coarsely hairy 6. **B.foetidum**

6 Corolla usually less than 25 mm long or in diameter:

10 Corolla tubular, campanulate or saucer-shaped and shallow:

11 Corolla-lobes 12-15 mm long, usually with margin recurved and appearing subterete:

12 Flower-bud slender, tapering; corolla-lobes dark purple, with margins recurved towards tips, often coarsely pubescent outside 6. **B.foetidum**

12 Flower-bud with terete beak; corolla-lobes greenish to yellowish, with margins recurved throughout:

13 Plant dwarf, tufted, up to about 50 mm high; pedicels slender, up to about 25 mm long . . . . 7. **B.montanum**

13 Plant up to about 150 mm tall, branched above; leaves undulate; flowers fasciculate; corona-pockets with long hairs within . . . . 8. **B.duplicatum**

11 Corolla-lobes usually less than 10 mm long:

14 Corolla-bud 6-7 mm long with terete or cylindric beak; lobes linear from ovate base, puberulous in lower half within; leaves strongly folded upwards; plant dwarf, with branches up to ± 100 mm high . . . . 56(b). **B.dimorphum** subsp. **gratum**

14 Corolla-bud with pyramidal or tapering beak; leaves not normally folded upwards:

15 Stems ascending, diffuse or sometimes becoming procumbent, not prostrate and elongating from the beginning:
16 Corolla 13-15 mm diameter when lobes expanded:
17 Corolla mainly maroon with yellowish blotches or bands (or yellowish with maroon):
18 Corona-pockets with vertical slit down outer margin; flowers often 2 or more from a node, opening together:
19 Corolla-lobes with long vibratile cilia and hairs within corolla-tube . . 9. **B.tuberosum**
19 Corolla-lobes without cilia and corolla-tube glabrous . . . . 10. **B.decipiens**
18 Corona-pockets with V- or U-shaped outer margin; flowers 1-2 from a node, opening successively:
20 Corolla-lobes with long, soft, caducous cilia and sometimes with long hairs on surface:
21 Tuber small, regular, producing 1-2 slender, erect, annual stems; outer margin of corona-pocket broadly U-shaped
16. **B.alpinum**
21 Tubers usually irregularly shaped, producing rhizomes and diffuse herbaceous branches; outer margin of corona pockets V- or narrowly U-shaped . . . . 17. **B.petraeum**
20 Corolla-lobes without long cilia but sometimes with inner surface pubescent:
22 Stems ± diffuse, slender and elongating; leaves mainly elliptic-lanceolate; corolla with transverse purple on yellow banding round corona
18. **B.australe**
22 Stems ± compact or spreading; leaves mainly broadly ovate to lanceolate; corolla-lobes mauve, blotched, with long hairs on upper surface . . . . 15. **B.remotum**
17 Corolla whitish in lower half with copious hairs; lobes with glabrous yellowish-green tips
23. **B.stellatum**
16 Corolla usually less than 10 mm diameter, mainly yellow-cream:
23 Corolla subrotate, bright clear yellow; pedicels slender, up to 37 mm long . . . . 22. **B.caffrum**
23 Corolla-lobe hardly spreading, less than 4 mm long, whitish, green- or brown-tipped 29. **B.delicatum**
15 Stems few to several, prostrate, elongate, corolla usually glabrous, yellowish- or maroon-banded:
24 Corolla 7-12 mm diameter, rarely larger, very shallowly saucer-shaped; sepals linear-lanceolate, usually projecting beyond sinuses; leaves noticeably diminishing in size towards tips of branches . . . . 19. **B.pulchellum**
24 Corolla 17-20 mm diameter, shallowly to broadly basin-shaped or campanulate . . . . 20. **B.coddii**

10 Corolla with flat base; lobes rotate, spreading horizontally from base; plants prostrate:
25 Corolla 7-10 mm diameter . . . . . . 19. **B.pulchellum**
25 Corolla 15-18 (-25) mm diameter:
26 Corolla surface glabrous . . . . . 21(a). **B.bruceae** subsp. **bruceae**
26 Corolla surface with long soft hairs . . 21(b). **B.bruceae** subsp. **hirsutum**
5 Corolla divided to about halfway to base:
27 Stem erect to somewhat diffuse and later possibly procumbent:
28 Corolla maroon-marked or maroon with yellowish bands or blotches, 10-20 mm diameter:
29 Corolla-lobes without long, vibratile, caducous cilia on margins:
30 Corolla glabrous within or minutely papillate:
31 Colour markings of corolla in circular or transverse pattern round corona, passing into veining directed to the tip of the lobes . . . . 13. **B.campanulatum**
31 Colour markings of corolla in circular pattern round corona, becoming dull red or maroon on lobes 14. **B.modestum**
30 Corolla pubescent on upper surface; lobes blotched 15. **B.remotum**
29 Corolla-lobes ciliate with long, vibratile, caducous hairs 16. **B.alpinum**
28 Corolla yellow and glabrous or whitish and hairy, usually less than 10 mm diameter; pedicels slender, up to 37 mm long:
32 Corolla campanulate, yellow, glabrous . . 22. **B.caffrum**
32 Corolla narrowly campanulate, whitish in basal half with long white or sometimes mauvish pubescence, tip of lobes yellowish green, glabrous . . . . 23. **B.stellatum**
27 Stems prostrate, elongate (corolla normally about two-thirds-divided) . 20. **B.coddii**
3 Inner corona-lobes ascending or incumbent, about twice as high as staminal column:
33 Corona forming 5 small pockets with outer margin slightly swollen between bases of inner lobes; inner lobes oblong, erect . . . 1. **B.meyerianum**
33 Corona forming 5 pockets with outer margin narrowly V-shaped or slit, sometimes with the shoulders slightly extended:
34 Corolla-lobes lanceolate to linear with long, irregular cilia; corolla-tube cylindric-campanulate, 6-8 mm diameter . . . . . 11. **B.meyeranum**
34 Corolla-lobes ovate-triangular with few long cilia towards sinuses: corolla-tube broadly campanulate, 18-20 mm across rim . 12. **B.cathcartense**
2 Corona 2-seriate, sometimes appearing 1-seriate by the reduction of the inner lobes, producing spreading or erect outer lobes or teeth or a campanulate or cylindric tube over-topping staminal column, or disciform; inner lobes various:
35 Corona not disciform:
36 Corona producing spreading or erect outer lobes or teeth:
37 Inner corona-lobes elongating above staminal column:
38 Corolla 12-22 mm long, spreading, 16-30 mm diameter:
39 Corolla-lobes lanceolate to linear with long, irregular cilia; corolla-tube cylindric-campanulate, 6-8 mm diameter . . 11. **B.meyerianum**
39 Corolla-lobes ovate-triangular with few long cilia towards sinuses;

corolla-tube broadly campanulate, 18-20 mm across rim . . . . .
12. **B.cathcartense**

38 Corolla ± 5 mm long, spreading, 6-9 mm diameter; corona producing outer lobes spreading to corolla tube, deeply bilobuled with the adjacent lobules curved and touching at the corolla-sinuses; inner corona-lobes incumbent-erect above the staminal column . . 31. **B.parvulum**

37 Inner corona-lobes incumbent on anthers or obsolete; not elongated above:

40 Corolla-lobes with inflexed tips, the bud ± truncate; outer corona-lobes deeply bilobulate with the lobules erect . . . 34. **B.tabularium**

40 Corolla-lobes sometimes slightly thickened at tips, not inflexed:

41 Corolla-lobes slightly opening to half spreading, not rotate or reflexed:

42 Stems erect, suberect or somewhat diffuse:

43 Pedicels about as long or longer than the petioles; corolla mainly pale-coloured with dark-tipped lobes:

44 Corolla 6-8 mm long; base tubular:

45 Corolla thinly pubescent on inner surface towards base of lobes, mainly white with brownish-tipped lobes; corona appearing 10-toothed . . . . 24. **B.thunbergii**

45 Corolla glabrous on inner surface, pale greenish with blackish-tipped lobes; corona with falcately spreading lobules
25. **B.praelongum**

44 Corolla 5 mm or less long:

46 Corolla-lobes longer than the tube:

47 Flowers 1-2 (-3) at nodes:

48 Corolla 3-4 mm long; base nearly flat:

49 Corolla yellowish, spotted with wine-red . . .
26. **B.occidentale**

49 Corolla white to pinkish with lobes green-tipped
27. **B.nanum**

48 Corolla ±2 mm long . . . 28. **B.micranthum**

47 Flowers few to several in fascicles at the nodes; corolla white with green-brown-tipped lobes . . . . . .
29. **B.delicatum**

46 Corolla-lobes shorter than tube (ex description) . . .
30. **B.schoenlandianum**

43 Pedicels conspicuously shorter than petioles and flowers often somewhat hidden by spathulate leaves; corolla maroon; outer corona-lobes thickened, deeply bilobuled with curved branches and few hairs . . . . 32. **B.minor**

42 Stems prostrate, elongate, corolla maroon; outer corona-lobes deeply lobuled with the branches curved behind inner corona-lobes . .
33. **B.swazicum**

41 Corolla-lobes rotate, spreading horizontally, becoming reflexed; corona exposed on corolla; outer corona-lobes emarginate, bifid, or deeply toothed:

50 Corona appearing equally 10-toothed, overtopping staminal column; inner corona-lobes incumbent on backs of anthers . . . . . .
35. **B.huttonii**

50 Corona outer lobes emarginate, bifid, or bilobuled, overtopping staminal column:

50a Corolla-lobes strongly reflexed, smooth on upper surface; flower stalks (pedicels) usually 10 mm or more long, slender; outer corona-lobes linear, bifid or emarginate, erect . . . 36. **B.arnotii**

50b Corolla-lobes spreading more or less horizontally, rugose on upper surface; flower stalks very short; outer corona-lobes subquadrate, emarginate, incurved . . 43. **B.dinteri**

36 Corona with outer lobes forming a campanulate or cylindric tube, equal to or overtopping staminal column:

51 Corona campanulate, sometimes with indentations opposite inner lobes and confluent from within with inner lobes; flowers 1 or 2(-3), developing in succession:

52 Stem erect, sparsely branched:

53 Corolla ±5 mm diameter, half to two-thirds divided, maroon; lobes reflexing to pedicel, with long white cilia on margins . . . . . . 37. **B.minimum**

53 Corolla ±15 mm diameter, ± half divided; tube broadly campanulate with circular greenish to almost black markings within, surface hairy . 38. **B.chlorozonum**

52 Stem diffuse to procumbent or prostrate:

54 Corolla yellow, 8-12 mm diameter; pedicels slender, up to 35 mm long; stem several from a tuber, diffuse . . . . 22. **B.caffrum**

54 Corolla not yellow:

55 Corolla with campanulate tube and spreading lobes, whitish within, becoming maroonish on lobes . . . . 39. **B.ngomense**

55 Corolla with short broad tube round base of corona; lobes rotate, black or brown-black with ring of long hairs round corona; leaves hairy, sometimes appearing silvery . . . 40. **B.incanum**

51 Corona cylindric, sometimes slightly angled, overtopping staminal column; flowers often in fascicles with more than one open at same time:

56 Corolla-lobes smooth on the inner surface:

57 Corolla-lobes mainly yellowish, rarely maroon, ovate-lanceolate; inner corona-lobes incumbent on anthers, outer slightly angled . . . . 41. **B.vahrmeijeri**

57 Corolla-lobes green, oblong-lanceolate, inner corona-lobes subobsolete, cushion-like at base of anthers . . . . . 42. **B.cupulatum**

56 Corolla-lobes rugose above, maroon, triangular; outer corona with slits opposite the pockets . . . . 43a. **B.brevipedicellatum**

35 Corona disciform, fleshy; inner corona obsolete, cushion-like at base of anthers; corolla rotate:

58 Corolla-lobes with long, subclavate, purple hairs; stem slender, erect or suberect; leaves linear, up to about 50 mm long . . . . . . 44. **B.schinzii**

58 Corolla-lobes glabrous; stem slender, suberect; leaves usually about as broad as long, up to ±15 mm long . . . . 45. **B.discoideum**

1 Corona fused to staminal column, expanding from near top, slightly spreading, with circumference usually with few fine teeth; corolla ±10 mm diameter, yellowish, usually with few medifixed hairs round base within; stems usually several, diffuse, sometimes flowering before development of leaves . . . . 46(b). **B.pygmaeum** subsp. **flavidum**

Group B. Rootstock producing a tuber; corolla-lobes united at the tips (sometimes becoming free in *B.pygmaeum, B.dimorphum* and *B.pachypodium*):

1 Corona fused to staminal column, expanding from near top, slightly spreading with circumference, usually with few fine teeth; corolla ±10 mm diameter, greenish yellow, usually with few medifixed hairs round base within; stems usually several, short, diffuse, sometimes flowering before development of leaves . . . . 46a. **B.pygmaeum** subsp. **pygmaeum**
1 Corona developing from or slightly above base of staminal column:
  2 Inner corona-lobes conspicuously elongated above staminal column:
    3 Corolla, with stout beak in bud, yellow, with short, campanulate tube; lobes sometimes becoming free, pinched-in at base; tubers irregular, sometimes partly exposed on rock-ledges, producing few annual branches, leaves broadly ovate to broadly elliptic-ovate . . . . 47. **B.pachypodium**
    3 Corolla with slender cylindric beak in bud, divided nearly to base; annual stems 1-2, slender:
      4 Corolla ±17 mm long, leaves linear . . . . 48. **B.gracile**
      4 Corolla ±45 mm long, leaves ovate, 10-15 mm long . 49. **B.gracillimum**
  2 Inner corona-lobes not elongated above staminal column:
    5 Flowers 1-few together at nodes:
      6 Outer corona almost obsolete to shortly campanulate or cupular or of 5 deeply divided bilobuled lobes:
        7 Corolla with few, long, cylindric-clavate, vibratile hairs within the corolla-tube near base of lobes . . . . 50. **B.gemmeum**
        7 Corolla without long cylindric-clavate, vibratile hairs within tube:
          8 Corolla with apiculate beak, becoming pyramidal when open; stems very slender, 50 mm or less high in the wild; flowers less than 5 mm long . . 51. **B.tenellum**
          8 Corolla with cylindric beak in bud; cage-like open part becoming broader than base:
            9 Plant covered with long shaggy hairs, at least on stems:
              10 Pedicels slender, 8-12 mm long; sepals linear-acuminate, 3-4 mm long; pubescence spreading, moderately dense; leaves often glabrous except for cilia . . . . 52. **B.tenue**
              10 Pedicels rarely more than 5 mm long; sepals less than 3 mm long; pubescence dense throughout:
                11 Corolla tube ±5 mm long; plant villous with spreading hairs . . 53. **B.villosum**
                11 Corolla tube 1-1,5 mm long or absent; plant pilose with hairs rather adpressed . . . . 54. **B.pilosum**
            9 Plant mainly shortly pubescent:
              12 Corolla-tube ±5 mm long; outer corona-lobes bifid or deeply bilobuled . . . . 55. **B.elongatum**
              12 Corolla-tube less than 3 mm long or absent:
                13 Outer corona-lobes almost obsolete or the outer margin very shortly bidentate; inner face of corolla-lobes puberulous in the lower half, outer mainly glabrous . . 56(a). **B.dimorphum** subsp. **dimorphum**
                13 Outer corona-lobes erect or suberect, usually overtopping staminal column, deeply bilobuled sometimes nearly to base; inner

face of corolla-lobes glabrous, outer usually pubescent . . . 57. **B.circinatum**

6 Outer corona cylindric-cupular, overtopping staminal column; corolla-tube cylindric, 5-7 mm long; lobes 9-14 mm long . . . . 58. **B.stenophyllum**

5 Flowers in compact terminal or subterminal, many-flowered heads on short peduncles; corolla-lobes 25-40 mm long . . . . 59. **B.barberae**

Group C. Rootstock producing a cluster of fleshy, cylindric or fusiform roots; corolla with little or no tube, lobes free at tips:

1 Stems more or less terete, pubescent; leaves ovate, oblong, broadly ovate or subcordate:

2 Corolla about 2 mm long; pedicels filiform, up to 20 mm long, exserted from leaves; outer corona-lobes shortly bifid, inner adpressed to base of anthers; leaves imbricate, subsessile, pilose, ciliate . . . . 60. **B.natalense**

2 Corolla 10-20 mm long:

3 Corolla glabrous outside; lobes constricted above staminal column, ciliate with long vibratile, sometimes clavate, purple hairs at the constriction:

4 Plant up to about 0,2 m tall; corolla glabrous except for cilia . . . . . . 61. **B.comptum**

4 Plant up to 0,9 m tall; corolla minutely puberulous on inner face . . . . . 62. **B.gerrardii**

3 Corolla pubescent on outer surface, sometimes thinly so, without long cilia:

5 Corolla-lobes hairy on both surfaces, about 20 mm long; pubescence coarse, brownish; outer corona-lobes, bilobuled; inner corona-lobes spathulate, oblanceolate, much exceeding staminal column . . . . 63. **B.macropetalum**

5 Corolla-lobes glabrous on inner surface, about 10 mm long; pubescence of plant short, smooth, greyish; outer corona campanulate; inner reduced to cushion-like swellings at base of anthers . . . . 64. **B.canum**

1 Stems angular or ribbed, glabrous or sometimes with few, short, straight hairs; corolla-lobes usually with papilla-like cilia:

6 Corolla-lobes less than 5 mm long:

7 Stem solitary, unbranched; corolla-lobes less than 2 mm long; outer corona-lobes bifid; inner corona-lobes connivent above the staminal column . . . . 65. **B.schizoglossoides**

7 Stems solitary or few, if solitary, then branched, corolla-lobes 2-4 mm long; outer corona 10-toothed:

8 Stems freely branched, up to about 0,15 m tall; inner corona-lobes not exceeding staminal column . . . . 66. **B.ramosissimum**

8 Stems moderately branched or more than one from a rootstock, 0,2-0,4 m tall; inner corona-lobes connivent, overtopping staminal column . . . . 67. **B.sandersonii**

6 Corolla-lobes 8-10 mm long; outer corona-lobes divided to middle or deeper; inner corona-lobes much overtopping the staminal column:

9 Corolla-lobes about 8 mm long; stems moderately branched; leaves rarely dense, rarely up to 30 mm long . . . . 68. **B.franksiae**

9 Corolla-lobes about 10 mm long; stems moderately branched, leafy at least on young growth; leaves linear, up to about 70 mm long . . 69. **B.longifolium**

SPECIES DESCRIPTIONS AND ILLUSTRATIONS

1. **Brachystelma blepharanthera** *H.Huber* in Mitt. bot. StSamml. München. 4:33 (1961); R.A.Dyer in Flora S.A. 27,4:7 (1980) [Fig.6]

Perennial dwarf herb with tuber. *Tuber* more or less 55 mm long and 25 mm broad producing a single slender annual stem. *Stem* sparsely branched up to about 150 mm high, minutely hairy. *Leaves* shortly stalked; blade 20-30 mm long, about 5 mm broad. *Flowers* 1-few together at nodes, two opening together or in succession. *Corolla* 7-9 mm long, divided to below middle, minutely hairy on outer surface; tube bell-shaped, 2-3 mm long; lobes free at tips.

Brachystelma blepharanthera, first described under the names *Blepharanthera dinteri* and *Blepharanthera edulis,* is only known from Damaraland in the northern area of S.W.Africa (Namibia). The old generic name *blepharanthera* was adopted for this species by Huber because both the names *B.dinteri* and *B.edulis* were already occupied for distinct species. Plants differ in their hairiness and in the size of the flowers but there seems no reason for a subdivision of the species on this account. An unusual feature, but not unique in the genus, is the tuber which is taller than broad. The 1-seriate corona within the base of the corolla is noteworthy. It shows no development of the outer series whilst the inner corona lobes are erect and conspicuous. In these respects it is readily distinguished from its dwarf allies.

It may be taken for granted that the tubers are edible and sought after by the indigenous peoples.

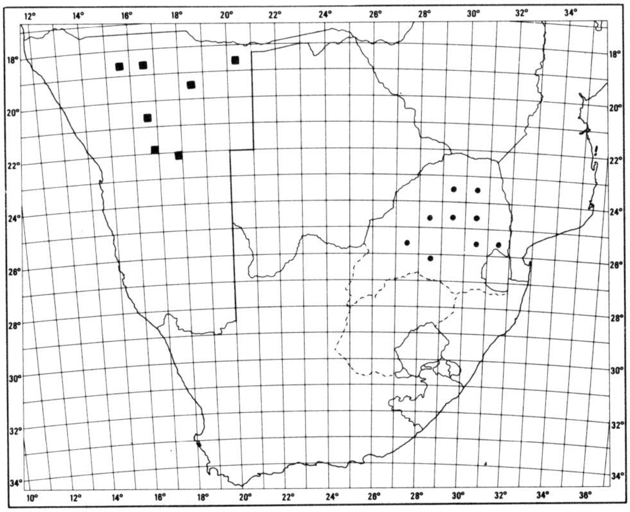

■ *Brachystelma blepharanthera* (1)
● *Brachystelma oianthum* (2)

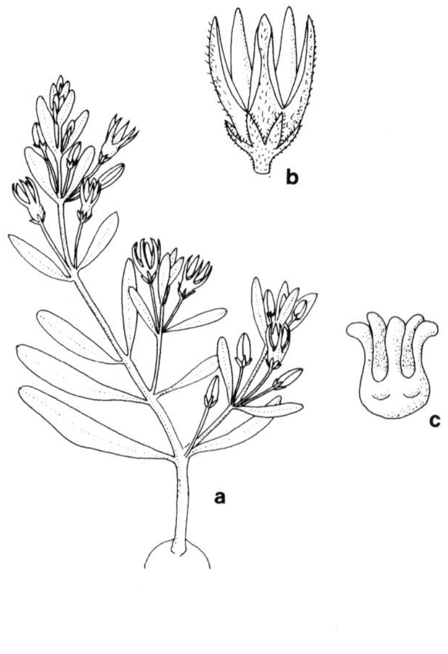

Figure 6. *Brachystelma blepharanthera:* a. habit, x ½; b. flower, x 2; c. corona, appearing uniseriate, outer consisting of thickened outer margin of nectariferous pockets and confluent with base of inner lobes; inner lobes elongate and considerably overtopping staminal column, x 5 (after Bot. Jb. 51:147 (1913))

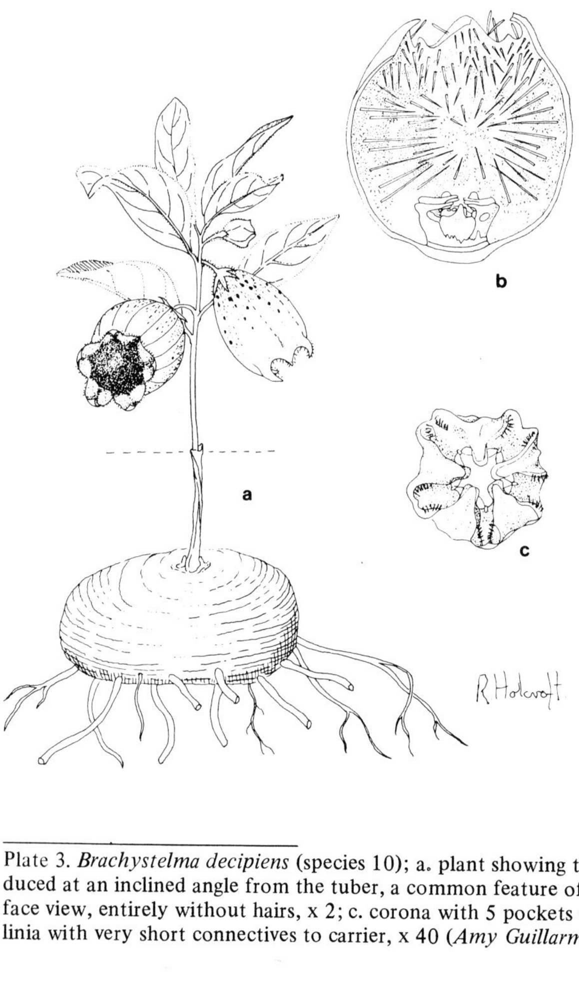

Figure 7. *Brachystelma oianthum:* a. habit, x ½; b. longitudinal section of flower, showing inner surface of corolla and exposing corona, x 1; c. top view of corona showing 5 pockets and incumbent inner lobes on staminal column (after Flower. Pl. S.Afr. Plate 537)

Plate 3. *Brachystelma decipiens* (species 10); a. plant showing the perennial stems to have been produced at an inclined angle from the tuber, a common feature of plants from that area, x 1; b. flower face view, entirely without hairs, x 2; c. corona with 5 pockets very similar to *B. tuberosum,* x 2; d. pollinia with very short connectives to carrier, x 40 (*Amy Guillarmod* 7501, near Grahamstown)

d
c
b
a
Auriol Batten

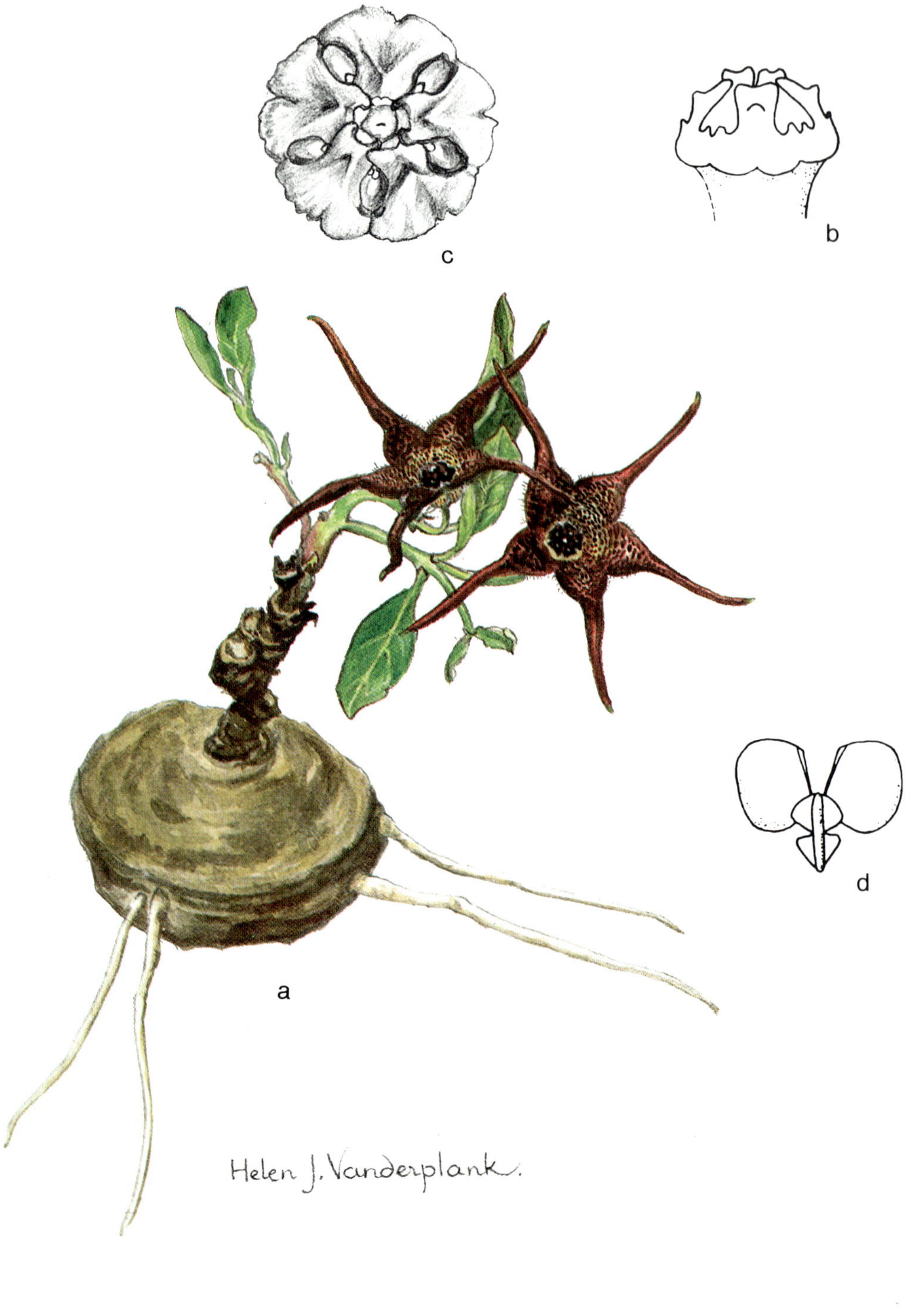
c
b
a
d
Helen J. Vanderplank.

Figure 7A. *Brachystelma oianthum:* plant in cult. BRI north of Nylstroom, Transvaal, in pockets of Waterberg sandstone, x 1 (*J. Vahrmeijer* 2475, photo: Adela Romanowski)

Figure 7B. *Brachystelma oianthum:* in flower and fruit in BRI, Nov. 1978, fruits (follicles) from previous season (*N.Jacobson,* The Downs, Letaba, N.Transvaal)

2. **Brachystelma oianthum** *Schltr.* in Bot. Jb. 20, Beibl. 51.53 (1895); R.A. Dyer in Flora S.Afr. 27,4:7 (1980) [Fig. 7, 7A, 7B]

Perennial dwarf herb with tuber. *Tuber* up to about 50 mm diameter, producing 1-few stems. *Stems* 1-2, slightly spreading, up to about 50 mm not often 80 mm tall, hairy. *Leaves* tapered into a short stalk; blade 10-60 mm long, 5-15 mm broad, minutely pubescent on both surfaces. *Flowers* solitary at the

---

Plate 4. *Brachystelma perditum* (species 5); a. plant, x 1; b. sideview of corona, x 5; c. corona from above showing 5 pockets of corona confluent with base of inner incumbent lobes round staminal column, the inner lobes with slight ridge from base ending in small horn about middle, truncate in upper margin round staminal column, x 6; d. pollinia attached to dumb-bell-shaped carrier, x 40 (*R.Bayliss* 7817, Rama Gate, Lesotho) nodes, spreading to nodding often under the leaves and resting on the ground, shortly stalked. *Corolla* 15-25 mm long, shortly 5-lobed; tube ovoid-campanulate, slightly narrowed to the mouth, yellowish outside with purple-brown mottling or entirely purple-brown, inside with long white hairs and minutely papillate; lobes free, with few clavate-vibratile, pale purple or rose-coloured hairs on margin.

Brachystelma oianthum, the species with an egg-shaped corolla, once recorded as *B.erianthum,* is reported to be locally frequent on mountains of the north-eastern Transvaal extending to south-western Transvaal and, according to N.E. Br. at Kew, into northern OFS, but I have not been able to confirm this last record. The flowers vary appreciably in size and colour pattern but there is no closely allied species with which it can be confused. It is in fact a connecting link between the two genera *Brachystelma* and *Ceropegia,* because of the relatively long, almost egg-shaped, corolla-tube and short free lobes. It would not have appeared very much out of place had it been transferred to *Ceropegia* and placed next to *C.pygmaea* and other dwarf species.

Judging by results at the Botanical Research Institute, Pretoria, the tubers are more amenable to cultivation than in most other species.

3. **Brachystelma caudatum** *(Thunb.) N.E.Br.* in J.Linn Soc., Bot.17:169 (1878); R.A.Dyer in Flora S.Afr. 27,4:8 (1980) [Fig. 8]

Perennial dwarf herb with tuber. *Tuber* up to 150 mm diameter, producing 1-several annual stems. *Stem* erect or spreading somewhat with age, branched from near base, up to about 100 mm high. *Leaves* tapered into a short petiole; blade linear-lanceolate to obovate or spathulate 15-30 mm long, up to 10 mm broad, usually with somewhat undulate margin, shortly hairy on both surfaces. *Flowers* 2-8 together, lateral at the nodes, shortly stalked. *Corolla* 30-45 mm long, hairy outside and on inner surface; tube 3-6 mm long, 4 mm diameter at mouth, spotted with purple-brown or black near mouth, yellowish towards base; lobes 20-40 mm long, ovate-lanceolate at base, tailed to tip with revolute margins, spreading-recurved, purple-brown spotted and greenish towards tips.

Brachystelma caudatum seems to have been given the name *caudatum* because of the slender tail-like corolla lobes. It was the first species of *Brachystelma* to be discovered but it was recorded under the name *Stapelia caudata* by the Swedish botanist Carl Thunberg in his Prodromus of 1790. This work is little more than a check list of the plants he gathered at the Cape during his travels from 1772-1775. The record was lost sight of under the name *Stapelia* with the result that most later collections of the same species were described under new names in the genus *Brachystelma, B.spathulatum* and *B.crispum.* The priority of the name *B.caudatum* was only re-established in the Flora of S.Afr. 1980. Although Thunberg may have collected his material on the Cape

Figure 8. *Brachystelma caudatum:* a. habit, x ½; b. corona appearing uniseriate, the outer forming 5 pockets with V-shaped outer margin and confluent with base of inner incumbent lobes, x 3 (after Plate 3016 in Curtis' Bot. Mag. 1830, under the name *B.crispum*)

Peninsula soon after his arrival in 1772, it was not until nearly 200 years later that it was possible to confirm its presence there, when in 1968 a single tuber in flower was unearthed on a site at Kommetjie near Simonstown being prepared for a coloured peoples housing scheme. It was brought to Kirstenbosch by P.Hattingh and identified by M.R.Mathews, curator of the living plant collection at Kirstenbosch at the time and who, on Harry Hall's advice, had good specimens prepared for permanent herbarium records. Flowers emitted a nauseous odour.

This species is recorded from the Cape Peninsula whence it extends to the Malmesbury, Darling and Ceres districts and may be also in Prince Albert and Caledon districts.

4. **Brachystelma glenense** *R.A.Dyer* in Bothalia 12,1:54 (1976); in Flora S.Afr. 27,4:9 (1980) [Fig.9]

Perennial dwarf herb with tuber. *Tuber* producing 1-2 annual stems. *Stems* branching from base, tufted, up to about 100 mm high, without hairs or with

Figure 9. *Brachystelma glenense:* dried type specimen in BRI collected at Glen College of Agriculture, OFS, in grassveld, x ±¾ (*J. W. Mosterdt,* 1952)

few minute ones. *Leaves* shortly stalked, linear-oblong up to about 30 mm long and 5 mm broad, hairless or with few hairs on undersurface which fall with age. *Flowers* 2-10 together, lateral at the nodes, opening more or less together. *Corolla* 35-40 mm long, without hairs on outer surface; tube campanulate, 4-5 mm long; lobes tapering from an ovate base, 30-35 mm long, margins recurved, shortly hairy towards base.

Brachystelma glenense is one of several species known only from one collection. It was discovered by Mr J.W.C.Mostert at the Glen Agricultural College, north of Bloemfontein, OFS. Mr Mostert was engaged in a botanical survey of grassveld camps when he discovered the one and only plant to come under his notice. In spite of subsequent diligent searching no further specimen was discovered. This is one of the many challenges to collectors to find more specimens in the neighbouring veld.

It has been stated that *B.glenense* is closely related to *B.caudatum* but appears to differ in size, hairiness and colouration. It is to enable a closer comparison that it is so important to discover more specimens in the wild. The photograph reproduced as figure 9 shows only the dried herbarium specimen but it should be a good guide.

5. **Brachystelma perditum** *R.A.Dyer* in Jl S.Afr.Bot. 43,1:10 (1977); in Flora S.Afr. 27,4:9 (1980) [Plate 4]

Perennial dwarf herb with tuber. *Tuber* 40-60 mm diameter, depressed, producing 1-few annual stems. *Stems* up to about 50 mm high, sparingly branched, shortly hairy. *Leaves* shortly stalked; the blade ovate-elliptic; 15-20 mm long, shortly hairy on undersurface and on margin. *Flowers* 2-several together, lateral at the nodes, sometimes more than one open at the same time. *Corolla* 18-25 mm long, brown mottled, divided ¾ to base; tube campanulate, 3-4,5 mm deep; lobes tapering from the base, with the margin recurving towards the tip, finely rugose and very finely papillate on upper surface, with or without long fine hairs near the base of the lobes.

Brachystelma perditum has a strange history. The name *perditum,* perdition, lost or destroyed, seemed appropriate when there appeared so little hope of the species being rediscovered in my time. Two records now exist, the first in Natal in 1907 near Ntabamhlope on the road to Giant's Castle in the Drakensberg and the second in 1976 in Lesotho north of Ramas Gate in the south Drakensberg near the Lesotho-Natal-Transkei boundary.

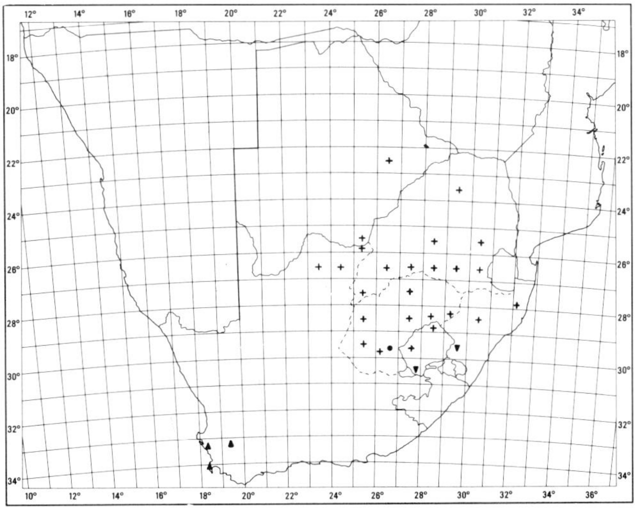

▲ *Brachystelma caudatum* (3)
● *Brachystelma glenense* (4)
▼ *Brachystelma perditum* (5)
\+ *Brachystelma foetidum* (6)

In the first place I was extremely diffident in describing a poor specimen which had lain nameless for 70 years in the Natal Herbarium. In addition its place of origin was uncertain. J.Wylie, curator of the botanic garden, Durban, made special expeditions to collect plant material for the Natal Herbarium under the directorship of J.Medley Wood. Wylie gave the locality for the *Brachystelma* merely as Niginya, 6 000 ft which must have seemed adequate at the time, but with the passage of time all trace of its whereabouts in Natal was lost.

It was only while the description was in press in 1977 that Mr Pierre du Toit, recently placed in charge of the Natal Herbarium for BRI, made a further search for possible clues to the correct locality. He established the fact, from other specimens in the herbarium, that in October 1907, Wylie was on a collecting expedition via Ntabamhlope (recorded as Tabamhlope) to Giant's Castle in the Drakensberg. A few miles north of the 'White Mountain' is Nyiginye, which is obviously what Wylie meant by Niginya.

In 1976 Roy Bayliss working from Grahamstown in the eastern Cape, made a collecting tour into the southern portion of Lesotho (Basutoland), neighbouring on Natal and Transkei. There in the mountains he came upon a small group of black Lesotho boys making a meal of plant tubers while they tended their sheep. Roy took a gamble and persuaded the boys to sell the uneaten tubers. These were cultivated in Grahamstown and flowered in December 1976, when it was realized that the plants were of special interest. It gave me great pleasure to confirm that they represented the second collection of *B.perditum.* The species has very distinctive subquadrate inner corona lobes which encircle the white centre of the top of the staminal column.

6. **Brachystelma foetidum** *Schltr.* in Bot.Jb.20, Beibl.51:52 (1895), R.A.Dyer in Flora S.Afr. 27,4:9 (1980) [Fig.10]

Perennial herb with tuber. *Tuber* 150-200 mm diameter, depressed from above, producing 1-several annual stems. *Stems* 70-150 mm tall or spreading, branching, variably hairy, often with short, rather coarse, spreading hairs. *Leaves* narrowed into a short stalk, linear-lanceolate to ovate or elliptic, 5-30 mm long, 3-10 mm broad, often somewhat folded upwards and undulate. *Flowers* 2-6 together, very variable in size. *Corolla* 25-50 mm diameter, tube campanulate, 6-10 mm long, white to yellowish within, with purple-brown warty spots, lobes 10-25 mm long, with or without hairs externally and usually without on inner surface, purple-brown to nearly black.

Brachystelma foeditum, whose flowers have the smell of putrid meat, is widespread in the grassveld of the Transvaal, Botswana, northern Cape, OFS and extends deep into Natal, over which distribution area it is particularly variable in size of flowers. N.E.Brown pointed out in Flora Capensis 70 years

Figure 10. *Brachystelma foetidum:* plant, x ½; this species has a wide distribution and the flower size varies considerably in relatively high to low rainfall areas; the tuber is coarser in texture than those of the more edible types (*Barkhuizen* 80, western Transvaal, cult. BRI)

ago that Schlechter founded *B.foetidum* upon a small-flowered specimen with the inner surface of the corolla-lobes without hairs and *B.rehmannii* upon a large-flowered specimen with the inner surface of the corolla-lobes shortly hairy (uncommon), but Brown found no structural difference between the forms nor did Schlechter mention any. There are many intermediate forms on record and the variability is attributed to environmental factors.

Whether the comparative abundance of the tubers has anything to do with the edibility of the tubers it is not possible to say for certain but it would appear so and one can guess that the flavour and texture are inferior.

7. **Brachystelma montanum** *R.A.Dyer* in Bothalia 6,3:540 (1956); in Flora S. Afr. 27,4:10 (1980) [Fig.11]

Perennial dwarf herb with tuber. *Tuber* 20-50 mm broad and up to 20 mm high, producing 1-few annual stems. *Stems* up to 60 mm high, erect or spread-

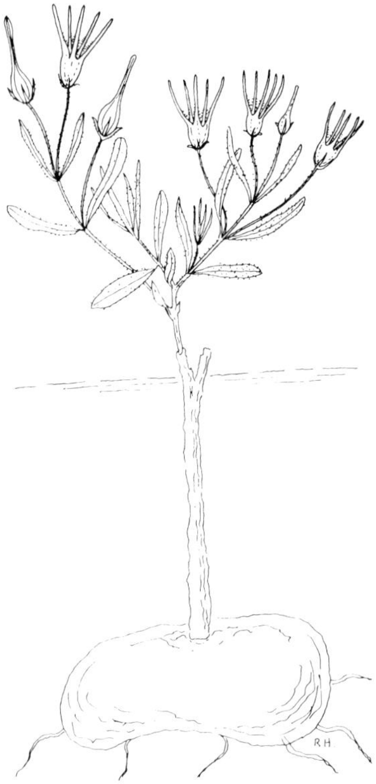

Figure 11. *Brachystelma montanum:* only known from the type specimen, x ½ (after *Galpin* 6278, near Queenstown, eastern Cape)

ing, branched, thinly pubescent. *Leaves* slender, 15-25 mm long, 2-4 mm broad, tapered to base, with hairs mainly along the margins. *Flowers* usually 2 together, lateral at the nodes, on slender stalks. *Corolla* up to 15 mm long; tube campanulate, up to 4,5 mm long, lobes slender about 10 mm long with recurved margins, ciliate below the middle with rather long hairs.

Brachystelma montanum, so named because of its habitat, is founded on only one record by E.E.Galpin who collected it on the Andriesberg near Queenstown over 75 years ago. It does not appear to have a close relative. This then is a challenge to the collector, who treasures rare plants, to climb the Andriesberg about November and search rocky outcrops which are favoured by Brachystelmas in general. The only illustration offered is a reconstructed drawing of the small preserved type at the Botanical Research Institute, Pretoria.

8. **Brachystelma duplicatum** *R.A.Dyer* in Bothalia 12,4:629 (1979); in Flora S.Afr. 27,4:10 (1980) [Fig.12]

Perennial herb with tuber. *Tuber* up to 80 or 90 mm long and up to nearly the same diameter with age, producing 1 annual stem. *Stem* rather stiff erect moderately branched up to 150 mm tall, puberulous. *Leaves* shortly stalked; blade oblong-lanceolate, up to 10 mm long, 3 mm broad, undulate, minutely papillose hairy on upper surface and sparsely hairy below, bluish-green. *Flowers* abundant, 2 or more lateral at the nodes in succession. *Corolla* 22-24 mm long with short tube; tube bell-shaped, 2-3 mm deep and 4 mm diameter, sometimes with a few long straight hairs within; lobes free at tip. 10-12 mm long, 2 mm broad at base, slender with revolute margins above, bright green with 3 mauvish veins near the base of each lobe and extending into the tube.

Brachystelma duplicatum was first described under the name *multiflorum* because of its many flowers but this was a duplication of an older name for a different species and necessitated the name change. So far it is known from the Fauresmith area of OFS and not recorded elsewhere. It is in karroid veld and sometimes associated with *Brachystelma circinatum* which has the corolla-lobes always united at the tips. The tuber of *B. duplicatum* is unlike the broad

Figure 12. *Brachystelma duplicatum:* collected near Fauresmith, OFS, rare, x 1 (collected and photographed by *J.A. van Zyl*)

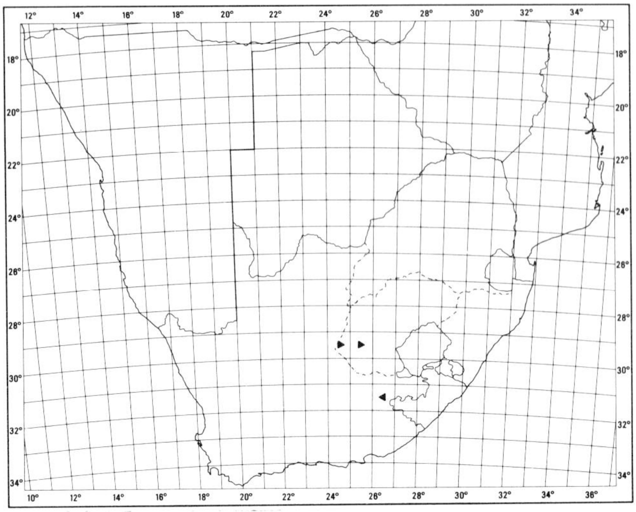

◀ *Brachystelma montanum* (7)
▶ *Brachystelma duplicatum* (8)

depressed pattern generally associated with *Brachystelma.* In *B.duplicatum,* it is elongated and soft in the young stage and only with age does it become more or less round but not depressed in the centre. Further records are highly desirable to establish its full distribution and range of variability.

9. **Brachystelma tuberosum** *R.Br.* in Curtis's bot.Mag.49:t.2343 (1822); R.A. Dyer in Flora S.Afr. 27,4:11 (1980) [Plate 5, Fig.13]

Perennial dwarf herb with tuber. *Tuber* up to about 50 mm diameter, depressed, producing 1-few annual stems. *Stems* up to about 100 mm tall, hairy. *Leaves* slender, rarely ovate to oblong, up to 35 mm long, 12 mm broad, tapered into a short stalk, usually more or less folded upwards, shortly hairy on undersurface and margins, without hairs on upper surface. *Flowers* 2-4, lateral at the nodes. *Corolla* ±20 mm long and in diameter; tube basin-shaped, 4-5 mm deep, without hairs on surface, green, speckled with purple-brown outside, yellow with transverse purple-brown line markings within; lobes lanceolate-linear, 6-7 mm long, with revolute margins, spreading-recurved, dark purple-brown, ciliate with long vibratile hairs on margin towards base.

Brachystelma tuberosum, as with many other species, has a tuberous root. It was described without exact information of its distribution but all evidence points to the southern Cape as the area from which the first tuber was introduced to England. The special importance of the species lies in the fact that it was the first to be given the generic name *Brachystelma* and all other species included in *Brachystelma* must be equated to it. Although it was first described

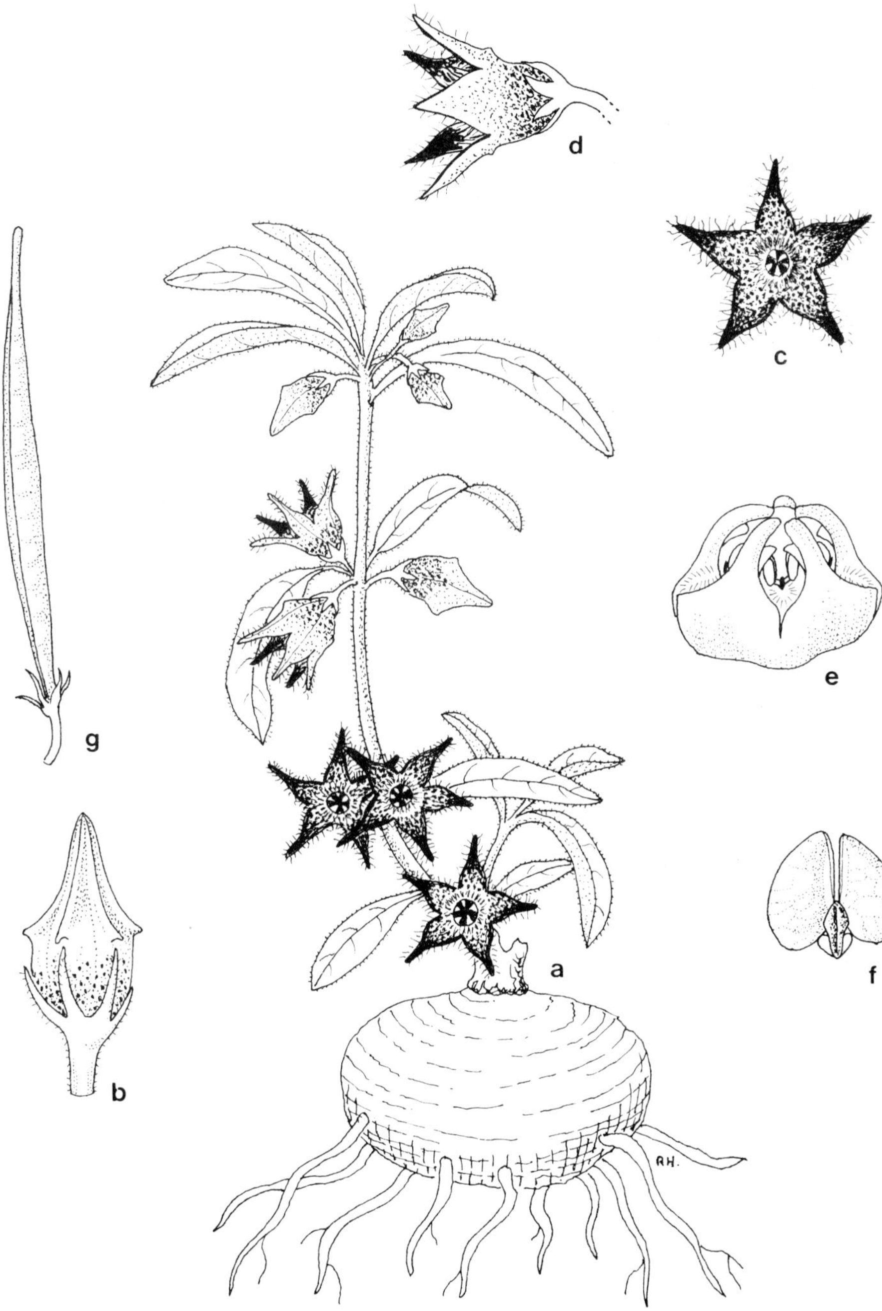

Figure 13. *Brachystelma tuberosum:* a. plant, x 1; b. bud showing spreading Huernia-like sinuses, x 2; c. flower face view, x 1½; d. side view, x 1½; e. corona, x 10; f. pollinia attached by short connectives to winged carrier, x 40; g. developing seed pod or follicle, x 1 (after Bot. Mag. t.2343, the illustration on which the genus *Brachystelma* is typified and *McLoughlin* in PRE 45101) (Figure 2 in Flora S.Afr. 27,4:(1980))

in 1822, it was not until 1953 that its presence in the Cape was confirmed when A.G.McLoughlin sent a flowering specimen from Knysna district to the BRI which corresponded sufficiently closely to the type illustration and description to be identified with it as *B.tuberosum*. In addition the Knysna district is a likely locality for the original collection because it lies on the route of the early botanical collectors making their way to the eastern Cape. Typical *B. tuberosum* seems to be absent from the coastal region from Humansdorp to Peddie but reappears in East London and Transkei. In the Port Elizabeth-Grahamstown section of the region, the closely allied *B.decipiens* is recorded, which differs in not having any long hairs on the corolla-lobes. This may be regarded as a trivial distinction by some.

In publishing the name *Brachystelma* in Curtis's Botanical Magazine 49: Plate 2343 (1822) John Sims, the editor, stated that he was highly indebted to the friendship of the learnéd botanist Robert Brown for enabling him to establish the genus of this very rare plant. In the synonymy Brown included the name *Stapelia tuberosa* Meerb. (Plant. Rar. t.54, f.1 (1789)) but Sims points out that Meerburgh's figure represents the flower much larger with the segments elongated to a very narrow point. Meerburgh's plant later proved to be equal to that described by Thunberg in 1894 as *Stapelia caudata,* now *Brachystelma caudatum* (Thunb.) N.E.Br. Thus R.Brown's inclusion of *Stapelia tuberosa* under his *Brachystelma tuberosum* must be regarded as a wrong identification.

10. **Brachystelma decipiens** *N.E.Br.* in Fl.Cap. 4,1:842 (1908); R.A.Dyer in Flora S.Afr. 27,4:11 (1980) [Plate 3]

Perennial dwarf herb with tuber. *Tuber* about 50 mm diameter, producing 1-few annual stems. *Stems* up to about 100 mm tall, sometimes arising from the tuber at an inclined angle, sparsely branched, with dense spreading or deflexed short hairs. *Leaves* narrow and tapering into the short stalk; more or less folded upwards, somewhat harshly shortly hairy on undersurface, without hairs on upper surface, minutely ciliate on margin. *Flowers* 2-4, lateral at the nodes. *Corolla* about 10 mm long, without hairs, yellow with transverse bands of purple-brown within and purple-brown on the lobes; tube bell-shaped, about 3 mm deep; lobes slender, spreading, somewhat twisted, about 7 mm long.

Brachystelma decipiens was so similar to *B.tuberosum* that N.E.Br. thought he was being deceived and gave it that name in consequence. Even now one wonders whether it should rather be regarded as a subspecies of *B.tuberosum.* Except for the lack of long hairs on the margin of the corolla-lobes and slightly more slender corolla-lobes there is little else to distinguish the two species but so far they have not been found to overlap in distribution in the wild. *B.decipiens* is described from the Port Elizabeth-Grahamstown area of the eastern Cape.

+ *Brachystelma tuberosum* (9)
● *Brachystelma decipiens* (10)
▲ *Brachystelma meyerianum* (11)
▼ *Brachystelma cathcartense* (12)

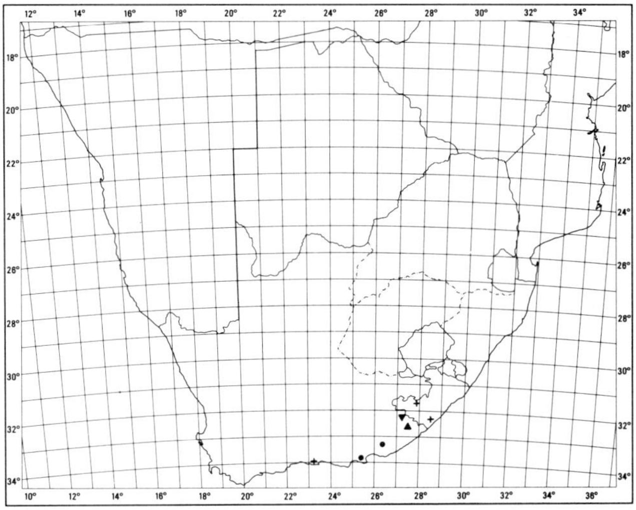

Collections of *B.decipiens* by Amy Jacot Guillarmod from the Zuurberg near Grahamstown, provided good material for illustration and were noted for the underground stems growing at an inclined angle to the ground surface which feature has not been seen with any other species.

11. **Brachystelma meyerianum** *Schltr.* in Bot.Jb. 21, Beibl.54:14 (1896); R.A. Dyer in Flora S.Afr. 27,4:13 (1980) [Plate 6]

Perennial dwarf herb with tuber. *Tuber* 40-60 mm diameter, producing 2-few annual stems. *Stems* branched from base, suberect or spreading up to 150 mm long, shortly hairy. *Leaves* slender up to about 30 mm long and less than 10 mm broad, variably hairy or sometimes without hairs. *Flowers* 1-4 together, lateral at nodes, shortly stalked. *Corolla* about 12 mm long, divided to half-way; tube more or less cup-shaped, 6-7 mm long, without hairs outside, pinkish to greyish-green or nearly white, and hairy within, upper part with long hairs; lobes tapering with somewhat reflexed margins, ciliate with long white hairs.

Brachystelma meyerianum is known only from the mountains of the eastern Cape from King William's Town to Komga into the Transkei. It was one of the earliest species to be recorded in the far eastern Cape having been discovered by J.F.Drège but was confused in the beginning with *B.tuberosum* R.Br. It was left to Schlechter to rename it much later after the German botanist C.A.Meyer who worked largely on the collections of Drège.

After many years of no records, botanical artist Auriol Batten re-established contact with this species near Stutterheim and noted considerable variability

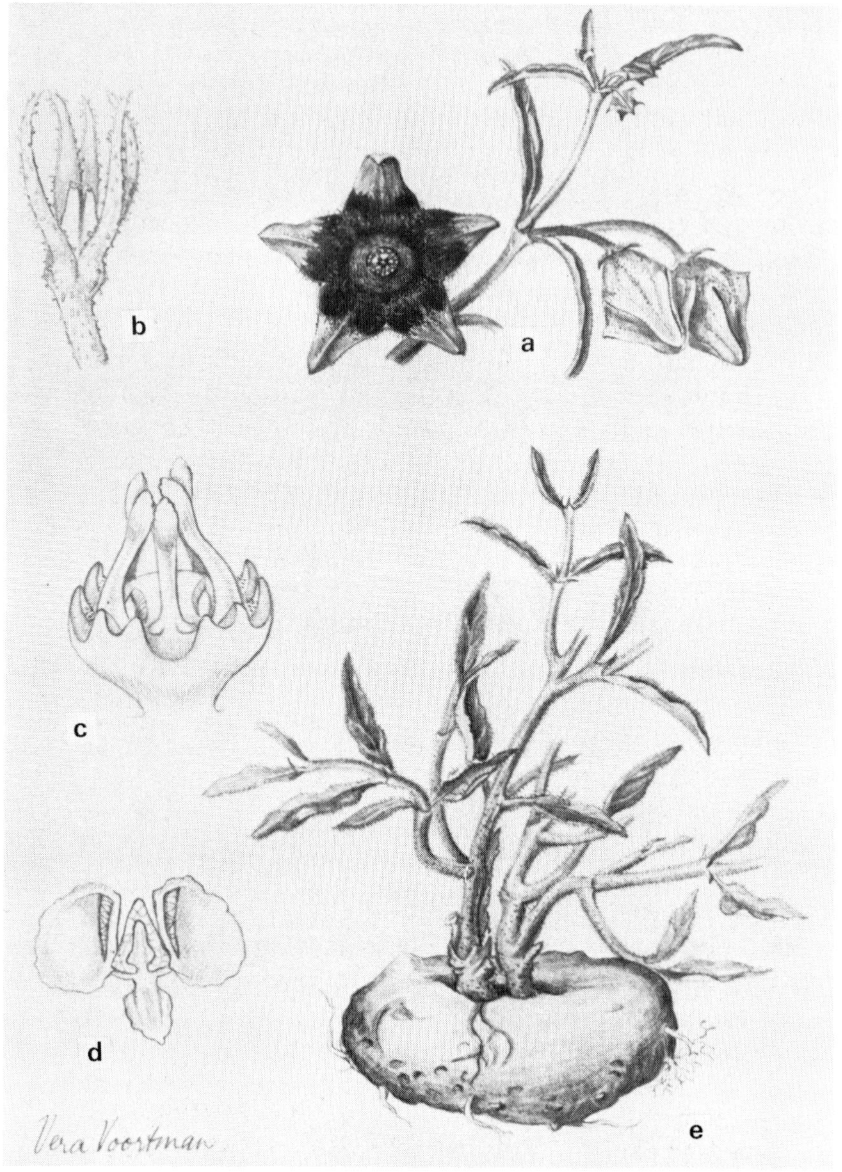

Figure 14. *Brachystelma cathcartense:* a. flowering twig, x 1; b. calyx with developing follicles, x 6; c. corona with bifid outer lobes and linear erect inner lobes, x 6; d. pollinia, x 75 (photo of plate 1667 in Flower. Pl. Afr. 1973)

in growth form and in the range of colour of flowers as indicated above. It was also of interest to note the appreciable difference between it and the comparatively recently described related species *Brachystelma cathcartense* with larger more open darker coloured and less hairy flowers.

12. **Brachystelma cathcartense** *R.A.Dyer* in Bothalia 10:431 (1971); Flora S. Afr. 27,4:13 (1980) [Fig.14]

Perennial dwarf herb with tuber. *Tuber* 40-50 mm diameter, usually depressed from above, rounded on base, producing 1-few annual stems. *Stems* sparsely branched from near base, normally 50-70 mm high. *Leaves* slender, up to about 25 mm long, tapered to base, minutely rough hairy, slightly folded upwards. *Flowers* 1-2 together, lateral at nodes, sometimes from opposite sides at the same time, with strong unpleasant odour. *Corolla* 20-22 mm long in bud, divided to about half-way, 10-ribbed, with 5 ribs extending to tips of lobes and others between with few or without hairs on outer surface; tube basin-shaped, nearly 10 mm deep, broadening somewhat abruptly at the mouth; finely maroon spotted on outside of tube, basal half within marked with radial maroon stripes and spots, becoming solid maroon and with long hairs in the upper half and on base of lobes; lobes broadly triangular, 13 mm long, with margin slightly recurved and tip incurved, shortly ciliate towards base with slightly swollen hairs.

Brachystelma cathcartense is known from the Cathcart district of the eastern Cape, hence its specific name. Flowering material was first collected by Mr P.F.du Toit, professional officer at the research station of Dohne, who took it to the Botanical Research Unit at Grahamstown, where Grace Britten diagnosed it as an undescribed species. It was arranged to have it illustrated by Vera Voortman and by degrees it has become well documented. It was established later that it had been collected by Harry Bolus as far back as 1896 and that John Acocks had found it in fruit in 1959. The most recent record was by Auriol Batten and party from East London who found it to be frequent in parts and grazed by sheep.

Within a relatively short period after the publication of the name, tubers were being advertised for sale in the nursery trade. Under cultivation, tubers may become considerably larger and more floriferous while the flowers retain their carion-like odour.

13. **Brachystelma campanulatum** *N.E.Br.* in Fl.Cap. 4,1:838 (1908); R.A.Dyer in Flora S.Afr. 27,4:14 (1980) [Fig.15]

Perennial dwarf herb with tuber. *Tuber* 50-75 mm diameter, depressed from above. *Stems* up to about 50 mm tall, usually unbranched, shortly hairy. *Leaves* 15-35 mm long, 10-15 mm broad, shortly hairy on both surfaces. *Flowers* solitary, nodding. *Corolla* bell-shaped, about 15 mm long, green and shortly hairy outside; tube about 10 mm long and same in diameter, with broken circular purple markings round corona, becoming greenish-yellow, spotted and longitudinally purple-brown veined into the lobes, with short

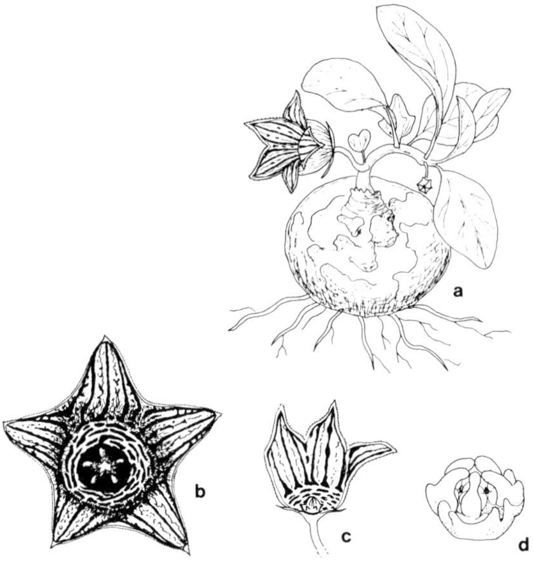

Figure 15. *Brachystelma campanulatum:* a. plant, x ½; b. face view of corolla showing markings on lobes directed to tips, x 1; c. longitudinal section of flower, with markings at base of tube in circular arrangement, slight magnification; d. corona, x 3 (adapted from Flower. Pl.Afr. t.1483A)

papillose-like hairs; lobes 4-5 mm long, broadly triangular, slightly spreading.

Brachystelma campanulatum was introduced to England by the British gardener-botanist Bowie in 1823 from near Bathurst at Linch's post in the eastern Cape. It was not until one and a half centuries later that Roy and Estelle Bayliss made the second record in 1965 and a third in 1970 in the same area of poor grassveld.

A good sketch of Bowie's plant is in the Kew Herbarium and it afforded N.E.Brown a clearer picture on which to draw up his description when naming the species with its campanula- or bell-shaped flower. A tuber sent to the Botanical Research Institute, Pretoria, by Roy and Estelle Baylisse in 1965 flowered here in September of that year and was the subject of the illustration which appears in Flowering Plants of Africa, 38: t.1483a (1967). The markings on the corolla are unusual in that the spots and bands are orientated towards the tips of the lobes, with only the basal markings being subcircular as is the usual pattern.

14. **Brachystelma modestum** *R.A.Dyer* in Flower. Pl.Afr. 30:t.1165a (1954); in Flora S.Afr. 27,4:15 (1980) [Fig.16, 16A]

Perennial dwarf herb with tuber. *Tuber* 20-30 mm diameter, somewhat depressed on upper side, rounded on underside. *Stems* 1-2 produced annually, rarely more, branched sparingly above, 20-30 mm high, slender, erect or suberect. *Leaves* in 3-5 pairs, shortly petiolate, 10-30 mm long, 5-10 mm broad, with or without hairs. *Flowers* 1-2 lateral at nodes developed successively. *Corolla* broadly bell-shaped, 80-90 mm long, 15 mm diameter, divided slightly less than half way; tube 2-3 mm deep, parchment-coloured within with subcircular broken dark red bands; lobes triangular, 4 mm long, velvety, maroon or chocolate-purple.

Brachystelma modestum, in its modest size, is one of a group of closely

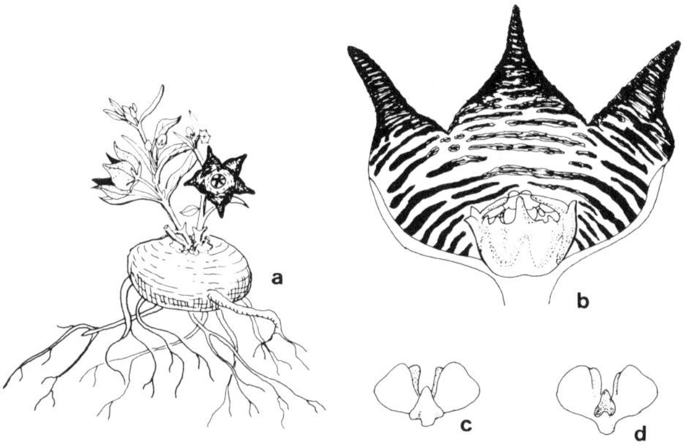

Figure 16. *Brachystelma modestum:* a. plant, x ½; b. longitudinal section of flower exposing corona, markings on corolla in circular pattern, x 3 (after Flower. Pl. Afr. 30.1165a)

Figure 16A. *Brachystelma modestum:* plant, with flower showing outer corona more divided than in type, otherwise similar, x 3 (*E.G.H.Oliver,* near the Kop of Kranzkop, Natal, cult. BRI, photo: Adela Romanowski)

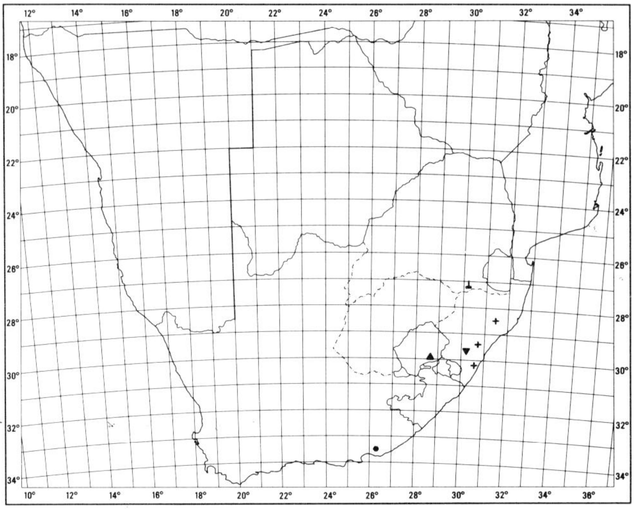

related species. In its typical diminutive upright form it is readily distinguished from its near allies, but records show that it does vary in habit over its range of distribution in the midlands of Natal. Under cultivation the deviation from normal is even greater.

The type was collected in Northern Natal near Nkandla and I have associated with it specimens from near Krantzkop and Noodsberg and from south of Durban near Umzinto which overlaps with *B.australe.* It has not been recorded from the Transvaal, but it has at least one close relative there in *B.coddii,* which, however, is always prostrate in habit or pendulous.

15. **Brachystelma remotum** *R.A.Dyer* in Bothalia 12,1:55 (1976); in Flora S. Afr. 27,4:15 (1980) [Fig.17]

Perennial dwarf herb producing a tuber. *Tuber* 25-40 mm diameter, somewhat depressed, sometimes partially exposed, producing 1-few annual stems. *Stems* sparsely branched from near base, spreading 30-60 mm long. *Leaves* up to about 15 mm long and broad, variably hairy. *Flowers* 1-2, opening in succession. *Corolla* 6-9 mm long, up to 18 mm diameter; tube broadly bell-shaped about 3 mm long, 6-7 mm broad; lobes more or less triangular, 4-7 mm long, 4 mm broad at base, with recurved margins, mauve or blotched, with long white hairs on inner surface.

Brachystelma remotum is so named because its distribution, which is rather off the beaten track on the mountains of northern Natal, is probably more common locally than the few known records imply. It is in the triangle of

Figure 17. *Brachystelma remotum:* plant from near type locality within the triangle Utrecht, Paulpietersburg and Wakkerstroom, in mountain shale outcrops, x 1 (*D.H.Roux,* PRE 41785)

Utrecht, Wakkerstroom and Paulpietersburg. To the south of this area one finds the related species *B.pulchellum, B.modestum* and *B.australe:* to the north in the Transvaal there are *B.bruceae* and *B.coddii.* Collections in eastern Transvaal by D.Hardy and J.Lavranos suggest that each isolated mountain habitat has a micro-climate which has caused local variants from a common stock.

16. **Brachystelma alpinum** *R.A.Dyer* in Jl S.Afr.Bot. 43,1:18 (1977); in Flora S.Afr. 27,4:15 (1980) [Fig.18]

Perennial dwarf herb with tuber. *Tuber* up to about 30 mm diameter, slightly depressed. *Stems* 1(2) produced annually, 30-50 mm tall (elongating in cultivation and becoming procumbent). *Leaves* shortly stalked, 10-20 mm long, 5-8 mm broad. *Flowers* 1(2) from 2nd or 3rd nodes of stem. *Corolla* 13-15 mm diameter, slightly rugose outside, divided to about half-way; tube broadly bell-shaped, about 2 mm deep, transversely banded with maroon; lobes triangular, 4-4,5 mm long, with margins slightly flexed towards tips, with sparse spreading long hairs on lower half and with few long, clavate, hairs at the base which fall with age.

Brachystelma alpinum is another discovery made by Roy Bayliss who came upon a group of these little plants near the high altitude Ramas Gate in South

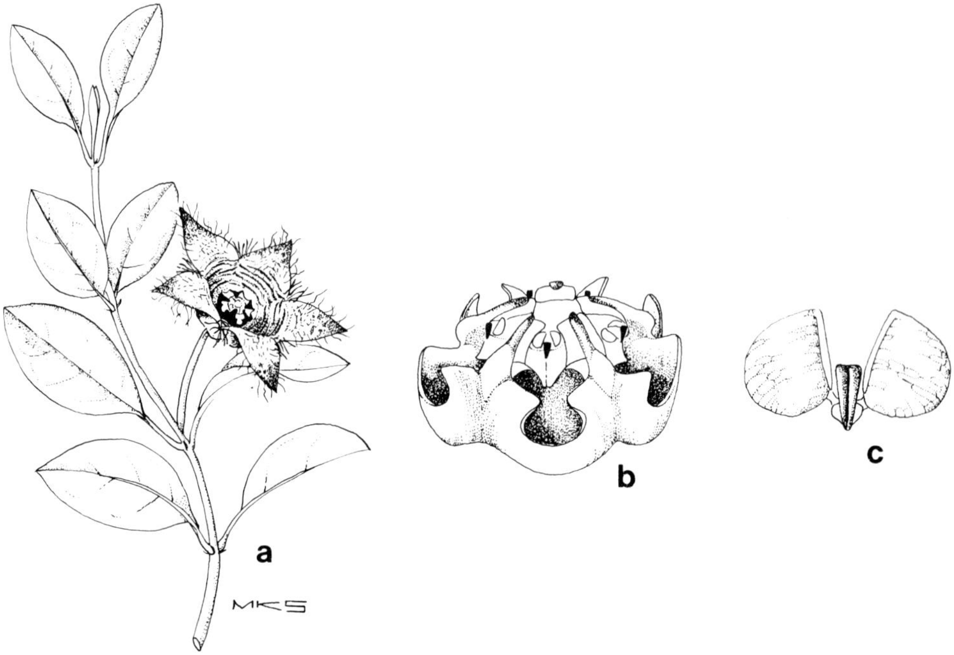

Figure 18. *Brachystelma alpinum:* a. stem in flower, x 1; b. corona with 5 pockets and broad U-shaped outer margin, confluent with base of incumbent inner lobes, x 5; c. pollinia shortly attached to winged carrier, x 25 (after *Bayliss* 819, Lesotho)

Lesotho in 1975. The tubers were at an altitude between 2 300-2 500 m in hard shale formation where the vegetation receives a high rainfall and experiences intense cold. It must be near the maximum limit of altitude for the genus.

The nearest related species appears to be *B.petraeum* from the foothills of the Drakensberg in Natal where the altitude is appreciably lower. The differences are best observed by a comparison of the illustrations where the hairs on the corollas and the structure of the coronas are of special significance.

17. **Brachystelma petraeum** *R.A.Dyer* in Jl S.Afr.Bot. 43,1:17 (1977); in Flora S.Afr. 27,4:16 (1980) [Fig.19, 19A]

Perennial dwarf herb with tuberous root-system. *Tubers* usually irregularly shaped, sometimes wedged between rocks and in fissures, sometimes producing a network of rhizomes in humus pads over flat rock surfaces. *Stems* produced annually, tufted or somewhat diffuse, branched, puberulous. *Leaves* 5-15 mm long, 3-8 mm broad, with or without few hairs. *Flowers* solitary on

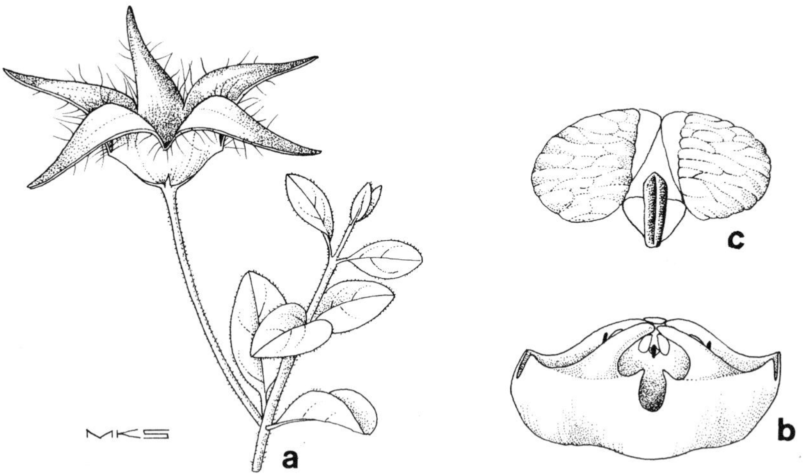

Figure 19. *Brachystelma petraeum:* branch in flower, x 2; b. corona with 5 pockets and outer margin narrowly U-shaped and confluent with incumbent inner lobes, x 10; c. pollinia attached to winged carrier, x 45 (after *Olive Hilliard* 5601, Richmond Natal)

slender stalk up to 20 mm long. *Corolla* divided about two-thirds to base, about 20 mm diameter; dull purple-brown outside, mottled-red or mustard-yellow within; tube shortly and broadly bell-shaped 3-5 mm long; lobes 7-10 mm long, tapering to apex, margin with long soft hairs mainly near base.

Brachystelma petraeum is known from the rocky escarpment above Byrne Valley near Richmond and further inland near Polela, Natal. Olive Hilliard discovered it first in Byrne Valley when looking for distribution records of *B. pulchellum.* It was found to be common locally at about 1 500 m altitude squeezed in rock crevices and it was this rock association which suggested the specific name. Later Joyce Stewart, also of the University of Natal, found it in the same area in humus pads over rocks and associated with *Selaginella* sp. and *Mariscus* sp. Within the humus pads the root-system gave rise to a network of tuberous rhizomes, a feature not seen to this degree in any other species of *Brachystelma* but approached in some of *Ceropegia.* A further record was made by Olive Hilliard and Bill Burtt at a higher altitude in the Underberg district where tubers were found in flat, damp, basalt gravel but they were not identified with certainty until they flowered at BRI, Pretoria, October 1979. The comparatively long erect flowering stalks attracted attention as also the long white rather crinkled hairs of the corolla.

Brachystelma petraeum belongs to a small group of related species among which the closest affinity is with *B.alpinum* and from which it is distinguished by the shape of the corona and distribution of the long crinkled hairs of the corolla.

Figure 19A. *Brachystelma petraeum:* showing a tuber on the left, right below showing the proliferation of a tuberous-rhizomatous rootsystem unlike anything seen in the genus before, top, mixture of rhizomes and aerial stems arising therefrom (*J.Stewart* 2022, Byrne Valley north of Richmond, Natal)

18. **Brachystelma australe** *R.A.Dyer* in Jl S.Afr.Bot. 43,1:12 (1977); in Flora S.Afr. 27,4:16 (1980) [Fig.20, Fig.20A]

Perennial herb with tuber. *Tuber* irregular in shape, sometimes becoming exposed above ground, with 1-few growing eyes. *Stems* few-several produced annually, somewhat tufted or elongating, especially in cultivation. *Leaves* 10-20 mm long, 3-8 mm broad. *Flowers* 1 from a flowering eye on slender stalk. *Corolla* about 20 mm diameter, broadly and shallowly basin-shaped, divided about two-thirds to base, with or mainly without hairs; tube 2-3 mm deep, transversely purplish blotched or banded on yellow background; lobes 8-10 mm long, about 4 mm broad at base, spreading-recurved with margins slightly reflexed.

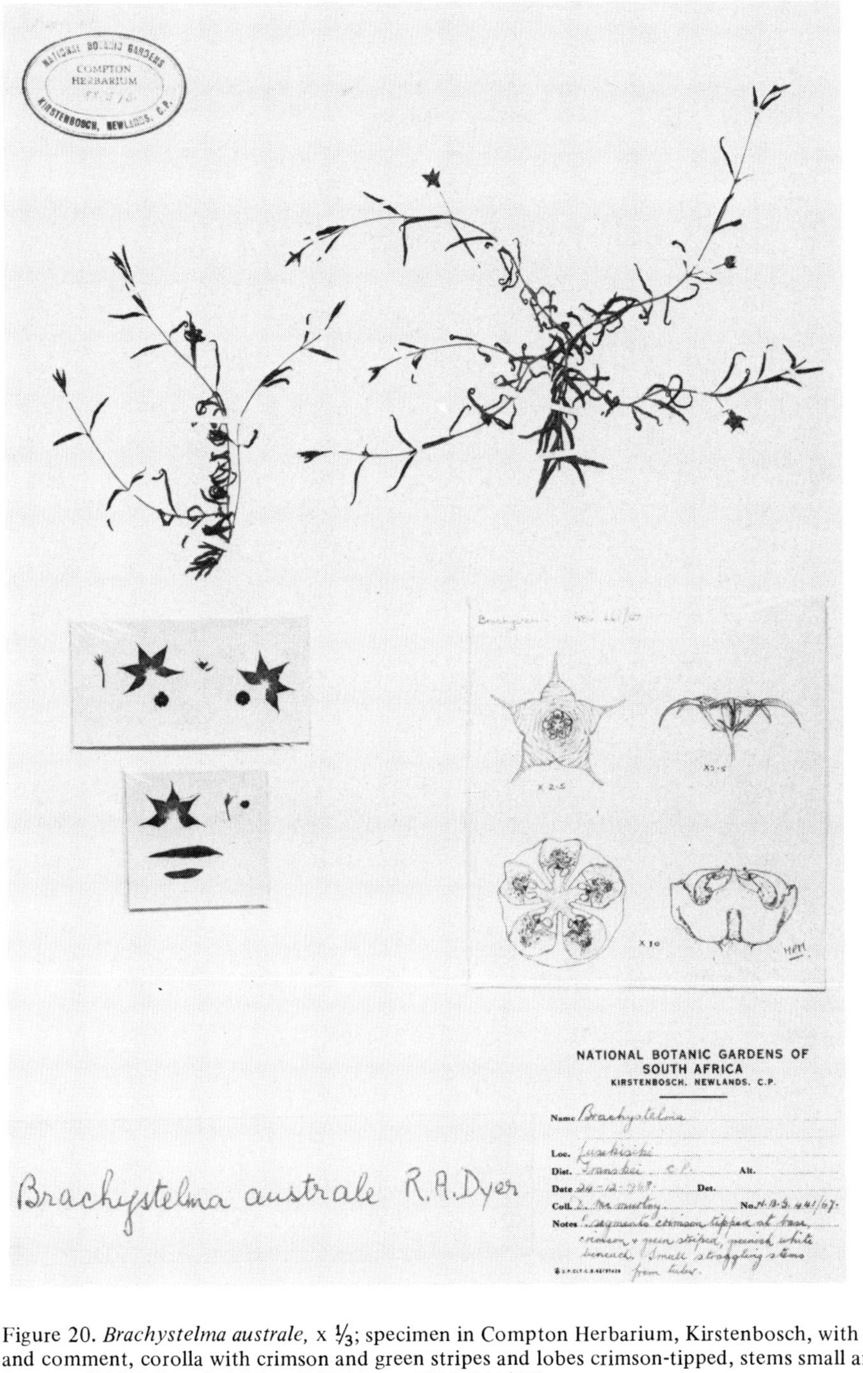

Figure 20. *Brachystelma australe,* x ⅓; specimen in Compton Herbarium, Kirstenbosch, with dissections and comment, corolla with crimson and green stripes and lobes crimson-tipped, stems small and straggling, from Lusikisiki, Transkei (after *D.McMurtry* NBG 441/67)

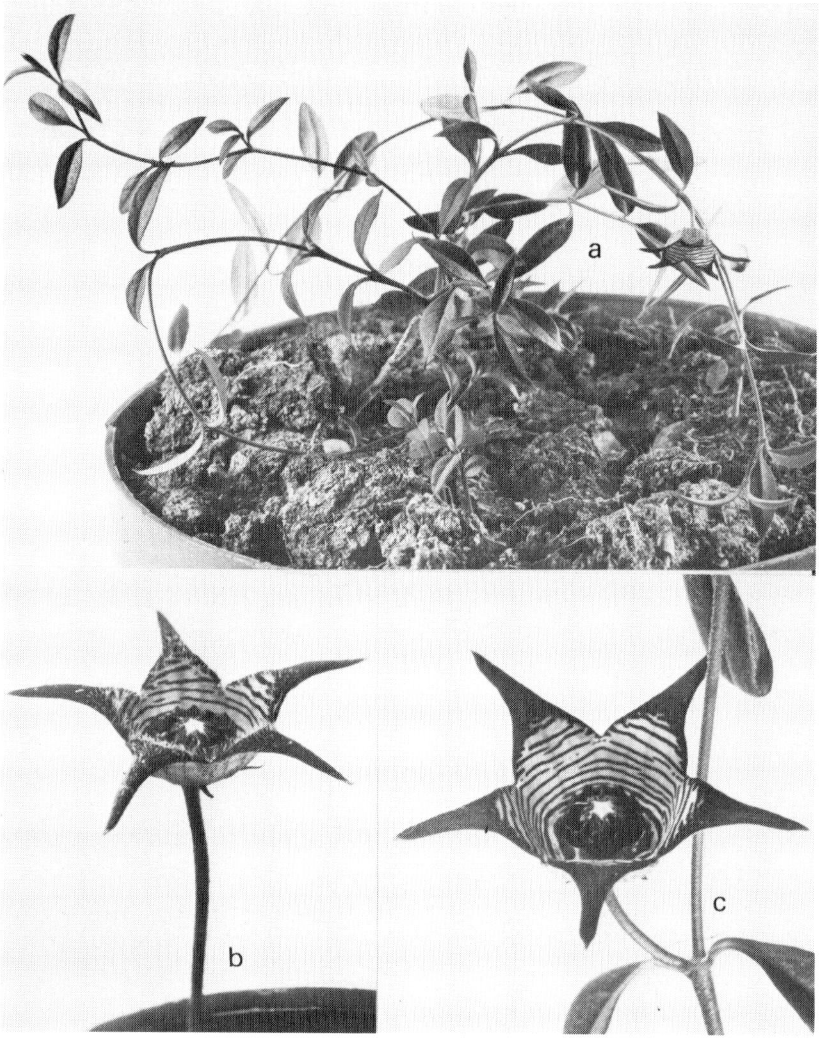

Figure 20A. *Brachystelma australe:* a. showing characteristic straggling growth, x 1; b. flower with hairs on corolla within, x 3; c. corolla without hairs, x 3 (after *S. Venter* 857 near Mkabathi Leper Institute, Pondoland, and other BRI records)

Brachystelma australe, named because of its southern distribution, has been recorded many times in the Pondoland coastal region growing in humus and on weathered sandstone ledges. In cultivation it develops a rather untidy tangle of branches but the flower retains some likeness to *B. modestum* and the far northern *B. coddii,* whereas its near neighbour in distance, *B. pulchellum,* remains distinctive in habit and size of corolla.

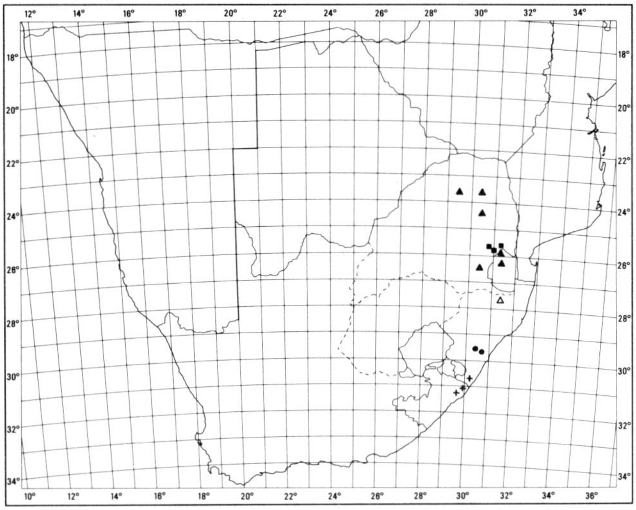

+ *Brachystelma australe* (18)
● *Brachystelma pulchellum* (19)
▲ *Brachystelma coddii* (20)
■ *Brachystelma bruceae* subsp. *bruceae* and subsp. *hirsutum* (21)

19. **Brachystelma pulchellum** *(Harv.) Schltr.* in Bot.Jb. 20, Beibl. 51:53 (1895); R.A.Dyer in Flora S.Afr. 27,4:17 (1980) [Plate 7, Fig.21]

Perennial herb with tuber. *Tuber* about 50 mm diameter, slightly depressed. *Stems* few-several produced annually, prostrate, up to about 25 cm long, with the size of the leaves decreasing in size as the season progresses. *Leaves* up to about 20 mm in early growth gradually diminishing to about 3 mm on the terminal portion. *Flowers* 1-2 on short stalks, opening in succession. *Corolla* 8-12 mm diameter (14 mm in cultivation), rotate, flat or with small depression containing basal portion of staminal column divided about half-way, maroon usually with transverse dull yellow banding round corona; lobes 2,5-4,0 mm long, becoming dark purple brown.

Brachystelma pulchellum, the beautiful one, was the first species described in a closely allied group. This makes it essential to fix its specific limits before recognising or creating additional species. Harvey described it as *Micraster pulchellus* in 1868 based on a specimen, *Sanderson* 342, from the midlands of Natal in short grassveld associated with outcrops of Table Mountain sandstone. Harvey likened the corolla to a miniature star-fish and exluded it from *Brachystelma* on a difference in coronal structure. *Micraster* and others of Harvey's splinter genera have since been absorbed under *Brachystelma.* One of the characteristic features of its truly prostrate branches is the gradual reduction in size of leaves and vigour to a thin tail giving the impression that the tuber had gradually exhausted its accumulated energy.

The specimen discovered by Sanderson has been matched by a number of more recent collections in similar situations between Durban and Pietermaritzburg. It does not, however, appear to stray much further, although a specimen from the Transvaal was figured under the name in Flowering Plants of Africa t.1121 (1952). This has since been reclassified as *B.bruceae* which is species 21.

Figure 21. *Brachystelma pulchellum:* a. tuber giving rise to few prostrate stems, x 1; b. corolla with very shallow or no tube, maroon with little or no marking, sepals (calyx segments) projecting beyond corolla sinuses, x 2 (*Hilliard & Burtt* 8590, Krantzkloof Nature Reserve, Natal)

20. **Brachystelma coddii** *R.A.Dyer* in Flower.Pl.Afr. 30:t.1181 (1955); in Flora S.Afr. 27,4:17 (1980) [Fig.22, 22A]

Perennial dwarf herb with tuber. *Tuber* 20-50 mm diameter, somewhat depressed. *Stems* 1-few produced annually, prostrate or sometimes pendant from

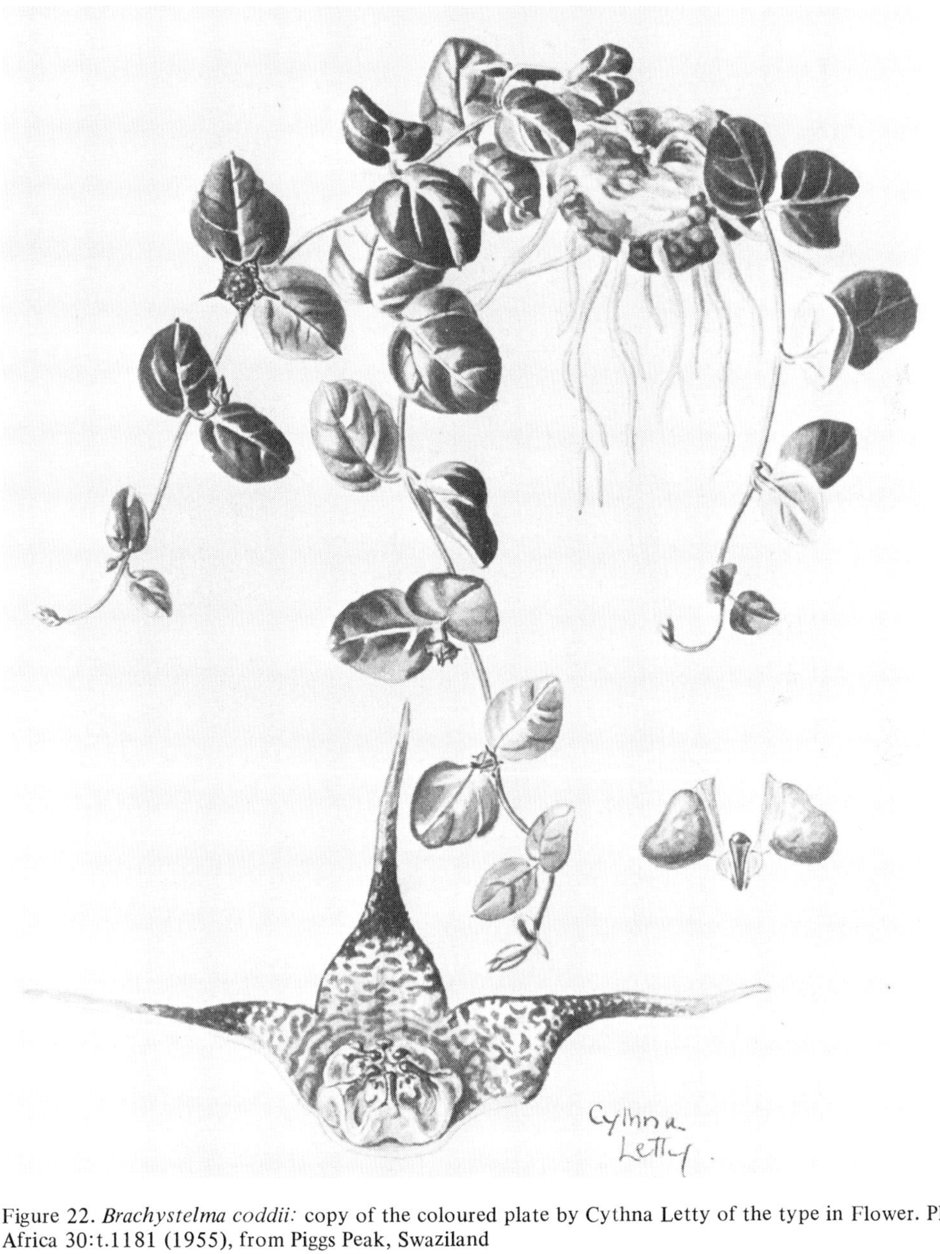

Figure 22. *Brachystelma coddii:* copy of the coloured plate by Cythna Letty of the type in Flower. Pl. Africa 30:t.1181 (1955), from Piggs Peak, Swaziland

Figure 22A. *Brachystelma coddii:* a. showing strongly prostrate and floriferous habit, corolla with basin-shaped base with or without markings, x ⅔; b. corona within corolla-basin and broad shortly beaked buds, x 2 (*J. Vahrmeijer* 2373, The Downs, northern Transvaal mist belt, sometimes pendulous on rock face, photo: Adela Romanowski)

rock ledges, up to about 30 mm long. *Leaves* mainly rounded and somewhat heart-shaped at base up to about 10 mm long (20 mm in cultivation) and progressively smaller as season develops. *Flowers* solitary or occasionally two from a node. *Corolla* about 10 mm long, 17-20 mm diameter, tube shallowly basin-shaped or saucer-shaped about 7 mm diameter, 3 mm deep, usually parchment coloured with flesh-coloured markings extending to base of lobes; lobes 6-7 mm long with margins somewhat recurved, velvety maroon, or scarlet to yellowish red, usually becoming-greenish yellow on tips but with a certain amount of variation.

Brachystelma coddii occurs on mountain sandstone outcrops of north-eastern Swaziland in the region of Piggs Peak, extending into high altitudes of the eastern Transvaal northwards towards Pietersburg and southwards to the vicinity of Carolina. Although typical forms of *B.coddii, B.bruceae* and *B.remotum* are distinctive, specimens from neighbouring mountains have caused some uncertainty as to their correct specific placing. Isolation seems to have given rise to distinctive local forms which call for intensive field study. In *B.coddii* which is named after Dr L.E.Codd, former director of BRI, one looks for a basin-like corolla-tube with circular markings, and enclosing the corona; the corona forms 5 pockets with the outer margin of each rather deeply V-shaped and forming shoulders to the confluent inner lobes.

B.coddii survives relatively well in cultivation and retains its prostrate character while flowering at the nodes for weeks on end. One collector near The Downs in northern Transvaal recorded branches dangling nearly a metre long from a rock ledge and also noted variations in the colouration of flowers and in the shape of leaves.

21. **Brachystelma bruceae** *R.A.Dyer* in Jl S.Afr.Bot. 43,1:13 (1977); in Flora S.Afr. 27,4:18 (1980) [Fig.23, 23A]

Perennial herb with tuber. *Tuber* up to about 70 mm diameter, sometimes developing tuberous outgrowths from the upper surface. *Stems* 1-few produced annually, branched near base, prostrate, up to about 120 (200) mm long, purple tinged. *Leaves* 10-15 mm long, 5-10 mm broad, dark purple-green above, paler on lower surface. *Flowers* solitary at the nodes. *Corolla* 1,5-1,8 mm diameter, divided two-thirds to four-fifths to base, opening more or less flat with the corona well exposed, purple-maroon or purplish with margins recurved.

Brachystelma bruceae from the eastern Transvaal was, for some years, without a specific name until E.A.Bruce identified it with *B.pulchellum* from Natal. With the accumulation of material and knowledge of the limits of *B.pulchellum,* the weight of evidence was in favour of elevating the Transvaal group to specific rank and separating it from *B.pulchellum.* This procedure gave the

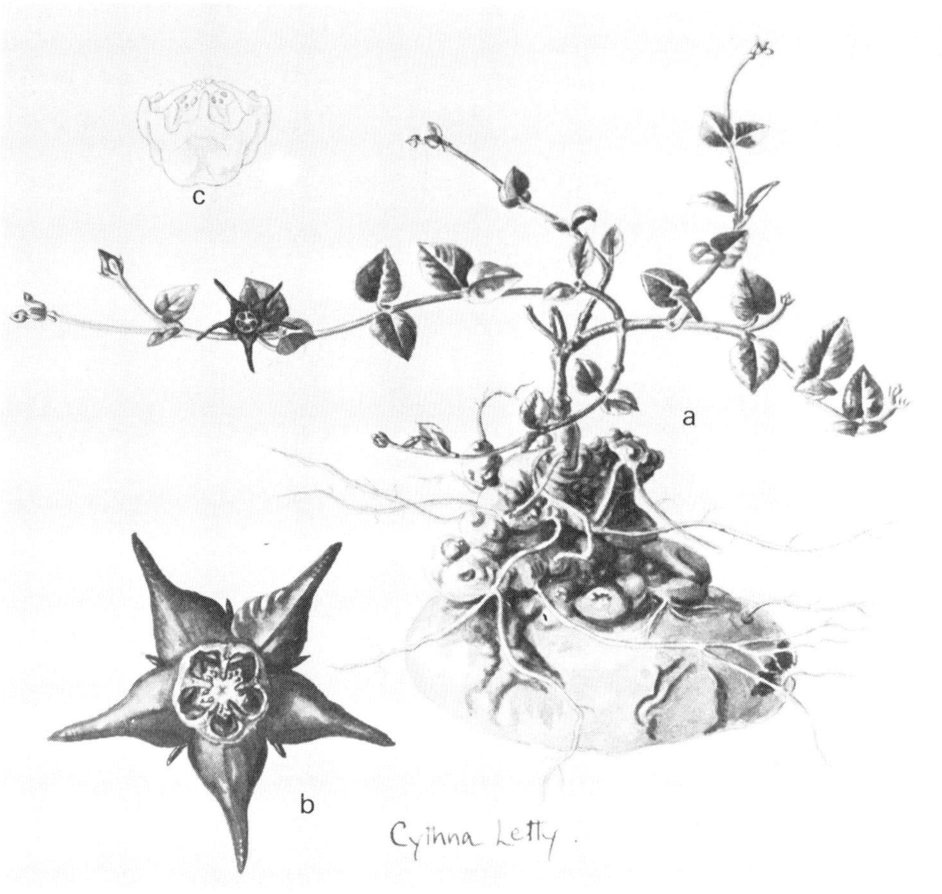

Figure 23. *Brachystelma bruceae* subsp. *bruceae:* a. plant with its tuber, x 1; b. flower, with spreading rotate corolla without tube, purple-maroon, x 4; c. corona with 5 deep pockets with outer margin notched, x 6, Kaapsche Hoop, eastern Transvaal, whitish sand over Black Reef Quartzite (after *L.E. Codd* in Flower. Pl. Afr. t.1121)

opportunity of commemorating Eileen Bruce who had devoted much of her time to the study of *Brachystelma* and *Ceropegia.*

Two subspecies are recognised and are distinguished as follows:

Corolla 15-18 mm diameter, divided two-thirds to three-quarters to base, purple-maroon, darker towards tips, shining, without hairs – subsp. **bruceae**

Corolla 16-18 (25) mm diameter, divided two-thirds to four-fifths to base, purplish with long soft purplish hairs on upper surface – subsp. **hirsutum**

subsp. **bruceae** [Fig.23]

The first specimen on record was collected in 1923 by J.H. & J.Davidson in the eastern Transvaal at Berlin, near Godwin River Station. Its main concen-

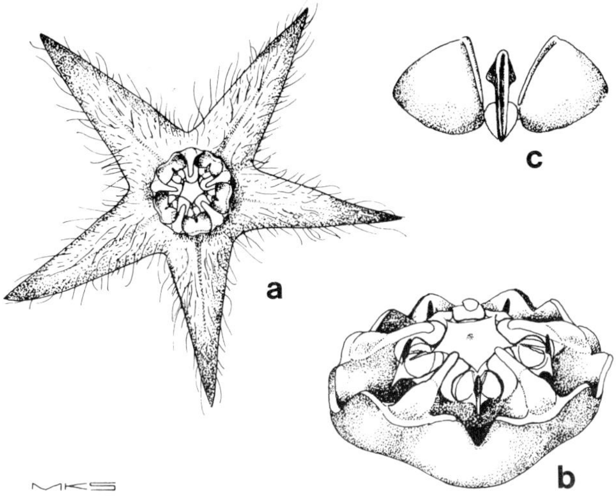

Figure 23A. *Brachystelma bruceae* subsp. *hirsutum:* a. corolla with long hairs on upper surface, x 3; b. corona with 5 deep pockets and outer margin with U-shaped notches, x 10; c. pollinia attached by short connectives (caudicles) to winged carrier, x 50 (after *Stirton* 1841 from near Barberton, Transvaal)

tration is on the Kaapsehoop mountains in sandy soil derived from Black Reef Quartzite.

subsp. **hirsutum** *R.A.Dyer* [Fig.23A]
The distinguishing feature of this subspecies is the presence of long soft purplish hairs on the upper surface of the corolla which averages a slightly greater diameter and the lobes are slightly longer and narrower than in subsp. *bruceae.* Although the distinguishing features of the subspecies are not great there is a habitat isolation, the one from the mountains and the other from the lowveld of the eastern Transvaal in the Barberton area.

22. **Brachystelma caffrum** *(Schltr.) N.E.Br.* in Gdnrs' Chron. 16:62 (1894); R.A.Dyer in Flora S.Afr. 27,4:19 (1980) [Fig.24]

Perennial dwarf herb with tuber. *Tuber* small depressed-globose. *Stems* several-numerous produced annually, diffusely spreading, becoming procumbent, 50-120 mm long, slender. *Leaves* shortly stalked, up to about 15 mm long and 9 mm broad. *Flowers* 1(2) lateral at nodes; stalked; stalks 15-30 mm long. *Corolla* 7,5-12,5 mm diameter, lobed to about two-thirds of its length, without distinct tube; lobes more or less triangular, 3-5 mm long.

Brachystelma caffrum is from that part of the eastern Cape known as Caffraria in early days and more specifically Ciskei. It was recorded in 1893 by T.R.Sim, then the local forester, on the Pirie Mt (now Mt Kemp) in the King William's Town district. He stated that it was abundant on rocky ground at

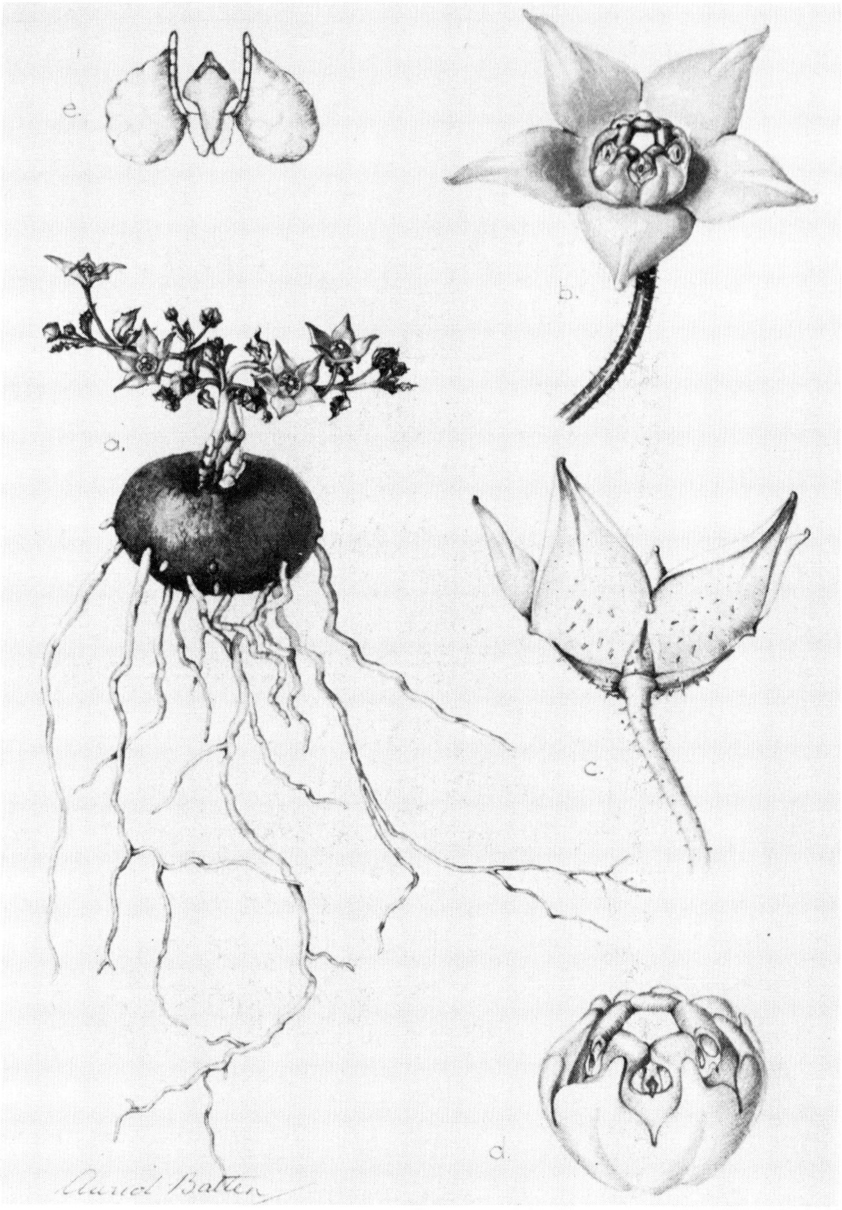

Figure 24. *Brachystelma caffrum,* photograph of coloured plate by Auriol Batten of a plant collected about 10 km south of Stutterheim on gravel outcrop, plentiful in that specific area: a. plant natural size flowered Sept. 1981; b. flower, x 5; c. back view, x 5; d. corona with fused lobes, x 10; e. pollinia, much enlarged

---

Plate 5. *Brachystelma tuberosum* (species 9); a. plant, x 1; b. flowering twig with open flower and bud, x 2; c. corona with outer margin of deep pockets, vertically slit with long hairs within, x 10; d. pollinia with short connectives to winged carrier, x ±60 (*Auriol Batten* 315, near East London, showing some variability from Figure 2 in Flora S.Afr. 27,4:12 (1980))

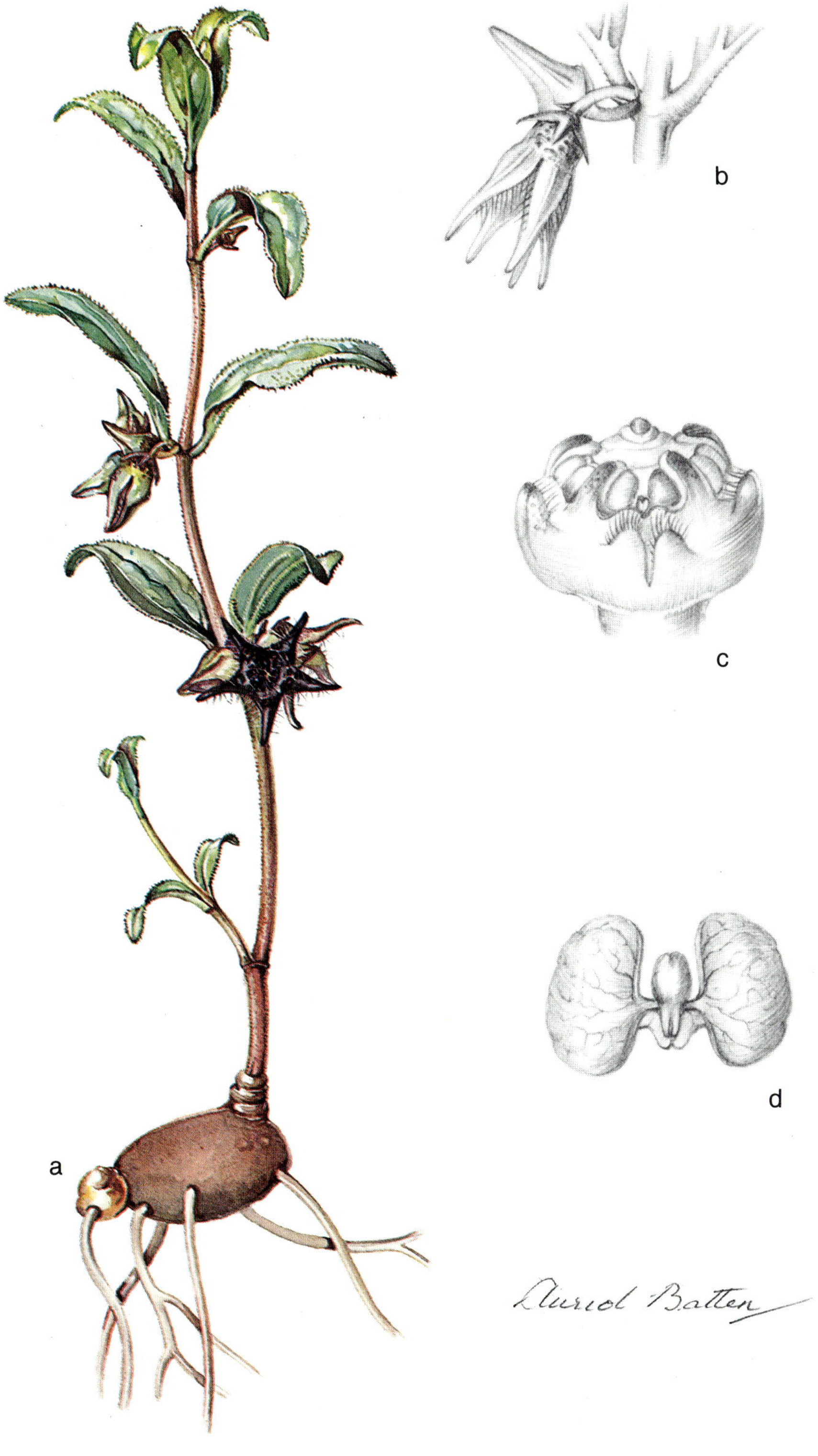
b
c
d
a
Auriol Batten

b
c
e
a
d
Auriol Batten

between 4 000-5 000 feet (1 200-1 500 m). It has not been recorded since. Auriol Batten and some botanical companions have scoured the area more than once in recent times without success. It seems improbable that the tubers have been eradicated entirely and the appeal must go out to all interested people to make a special effort to relocate the species. Until fresh specimens are found it will not be possible to improve on the diagrammatic illustrations on the herbarium sheet in the Bolus Herbarium, University of Cape Town. Some other herbaria have material from Sim's original collection or collections under his number 315. Schlechter was quick to describe it under the name *Tapeinostelma caffrum* but in the following year, 1894, N.E.Brown transferred it to *Brachystelma,* using a specimen flowering at the Royal Botanic Gardens, Kew, for his description. Brown remarked that the flowers stand up well above the trailing stems and although not large are cheerful and attractive so long as the nose is kept away from them, for their odour is by no means agreeable.

As we are in press, material collected by Auriol Batten and Peter Bruyns early in 1981 about 10 km south of Stutterheim on a rocky outcrop, has flowered in East London and BRI confirming the continued presence in quantity of typical *B.caffrum* in Sim's original stamping ground. Of the tubers sent to BRI by Peter Bruyns, however, under his no.2137, one turns out to be *B. meyerianum,* a related similar species which Auriol Batten had also collected near Stutterheim and illustrated in 1977. This confirms the similarity of the tubers and their similar environment and emphasizes the necessity of examining flowers before venturing a specific identification. The small bright yellow flowers of *B.caffrum* are unusual in the genus. Auriol Batten records that they smell like cow-droppings, and that the leaves showed great variation in size and shape, many having fluted margins.

23. **Brachystelma stellatum** *Bruce & Dyer* in Flower.Pl.Afr.30:t.1165B (1954); R.A.Dyer in Flora S.Afr. 27,4:19 (1980) [Fig.25, 25A]

Perennial dwarf herb with tuber. *Tuber* subglobose, 20-50 mm diameter compressed from above, rounded below. *Stems* few-several annually, branched from near base, suberect to spreading. *Leaves* shortly stalked, 4-5 pairs to a stem, up to about 10 mm long, 5-10 mm broad. *Flowers* 1-2 lateral at nodes on slender stalks 15-20 mm long. *Corolla* about 5 mm long, opening to about 10 mm diameter; tube containing corona, 2-2,5 mm deep, 2 mm diameter, cream tinged with purple and with purplish hairs; lobes spreading abruptly,

---

Plate 6. *Brachystelma meyerianum* (species 11); a. plant, x 1; b. twig with flower showing colour variation, x 1; c. flower with bell-shaped corolla-tube, x 4; d. corona shortly basin shaped at base forming 5 deep pockets with narrowly V-shaped outer margin and confluent with base of inner lobes, inner lobes incumbent-erect much over-topping staminal column, x 10; e. pollinia attached to winged carrier, x 40 (*Auriol Batten,* near Stutterheim)

Figure 25. *Brachystelma stellatum:* plant from Steenkampsberg in the Lydenburg district of the eastern Transvaal, showing diffuse floriferous growth, with long pedicels, x 1 (*Codd* 8050, photo: Harry King)

more or less triangular, with long white hairs, whitish to yellowish-green on tip.

Brachystelma stellatum may be looked upon as the counterpart in the Transvaal of *B.caffrum* in the eastern Cape. Both species are characterised by an abundance of small star-shaped (stellate) flowers on relatively long slender stalks and it is the hairiness of the corolla of *B.stellatum* by which it is readily distinguished. It occurs on the mountains of the northern Transvaal, East and

Figure 25A. *Brachystelma stellatum,* plant from Mt Anderson in eastern Transvaal: a. showing the corona closely surrounded by the corolla tube and the densely hairy corolla lobes, x 3; b. densely floriferous plant in cultivation, BRI, x 2 (*J.Lavranos,* photo: Adela Romanowski)

West of Lydenburg and originally found by L.E.Codd and B.de Winter in pockets of whitish sandy soil over Black Reef Quartzite at about 2 150 m altitude. *B.stellatum* has several times been recorded as frequent under similar conditions in the Lydenburg mountains and over the border of Pilgrims Rest.

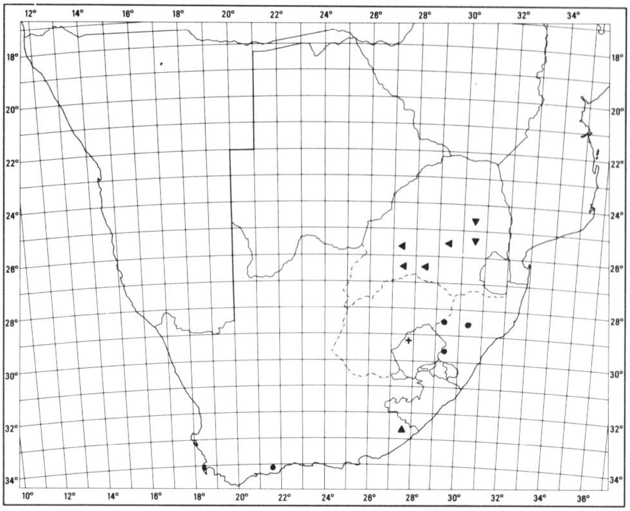

▲ *Brachystelma caffrum* (22)
▼ *Brachystelma stellatum* (23)
● *Brachystelma thunbergii* (24)
\+ *Brachystelma praelongum* (25)
■ *Brachystelma occidentale* (26)
◀ *Brachystelma nanum* (27)

24. **Brachystelma thunbergii** *N.E.Br.*, in Flora Cap. 4,1:843 (1908); R.A.Dyer in Flora S.Afr. 27,4:19 (1980) [Fig.26]

Perennial dwarf herb with tuber. *Tuber* up to about 40 mm diameter depressed. *Stems* 1-few produced annually, 50-100 mm tall, suberect, sparingly branched. *Leaves* up to about 40 mm long, 12 mm broad tapering into a short stalk. *Flowers* (1)2-8, lateral at the nodes. *Corolla* up to about 10 mm long; tube up to 2 mm long, 3-5 mm diameter at mouth, bell-shaped, white, thinly hairy within upper part; lobes up to 6 mm long, slender with recurved margins and incurved tip, thinly hairy on inner surface, greenish towards base, brownish or tawny-yellow towards tips.

Brachystelma thunbergii is named after the famous Swedish botanist Carl Peter Thunberg, often referred to as the father of South African botany. The species has an interesting history of discovery and distribution. Thunberg first collected it at the Cape a year or two after his arrival in 1772, but he classified it under the genus *Cynanchum.* It was not until N.E.Brown studied the material for his revisions in Flora Capensis 1908 that he placed it correctly under *Brachystelma.* A few more records have been made in the south-western Cape from Malmesbury to Caledon and Riversdale districts. From here there is an extraordinary jump to the foothills of the Natal Drakensberg in the Underberg district. It has been recorded at least three times in this environment and quite recently S.Venter found a colony on Jobskop, an isolated mountain which, as the crow flies, lies between Ladysmith and Pomeroy in northern Natal. When these plants flowered at BRI it was noted that they had a sweet honey scent, rather unusual in the genus and recalling the observation of E.E.Galpin

Figure 26. *Brachystelma thunbergii:* a. floriferous plant in cultivation at BRI from Jobskop between Ladysmith and Pomeroy in Natal, x ½; b. plant with flowers and developing seed pods or follicles, the follicles develop in pairs from a single flower and are sometimes referred to as buck-horns, x 1; c. enlarged flowers to illustrate depth of tube and slightly inflexed lobes, x 1½ (*S. Venter* 1173, photo: Adela Romanowski)

concerning *B.villosum (53)* and in contrast to *B.foetidum* and to N.E.Brown's view of *B.caffrum.*

25. **Brachystelma praelongum** *S.Moore* in J.Bot., Lond. 40:384 (1902); R.A. Dyer in Flora S.Afr. 27,4:20 (1980) [Fig.27]

Perennial dwarf herb with tuber. *Tuber* 30-40 mm diameter. *Stems* 1-2 produced annually, sparsely branched, slender 80-150 mm high. *Leaves* narrow, up to about 30 mm long, with infolded margins. *Flowers* 1-2 lateral at the nodes on slender stalks 15-45 mm long. *Corolla* about 7 mm long; tube cup-shaped, apparently slightly 5-ribbed outside, greenish-yellow; lobes free at tips, about 5 mm long, with recurved margins, hairless with dark coloured tips.

Brachystelma praelongum seems to be restricted in its distribution to a limited area of the north-eastern Orange Free State and northern Lesotho. Moore, who named and described the species, considered that the dwarf plant had unusually long flowering stalks for its size. It had been collected by Capt. Barrett-Hamilton who served in the South African war in 1901 in the Royal Irish Rifles. Some few years later Madam Dieterlen gathered more than one dwarf species of *Brachystelma* near Leribi, Lesotho, but only the one identified as *B.praelongum* was in flower. This species is closely related to *B.thunbergii* and it would be reassuring if living material could be studied to show that the lack of hairs on the inner surface of the corolla of *B.praelongum* and differences in colour and coronal structure were sufficient cause for its specific rank.

26. **Brachystelma occidentale** *Schltr.* in Verh.bot.Ver.Prov.Brandenb. 35:53 (1893); R.A.Dyer in Flora S.Afr. 27,4:20 (1980) [Fig.28]

Perennial dwarf herb with tuber. *Tuber* small subglobose, unlikely more than about 20 mm diameter. *Stem* solitary or perhaps 2, delicate, produced annually up to about 80 mm tall. *Leaves* narrow up to about 15 mm long tapering into a short stalk. *Flowers* 1-2, lateral at the nodes on delicate stalks about 5 mm long. *Corolla* about 3 mm long, yellowish-white, dotted with wine red on inner surface; tube about 1 mm long, cup-shaped; lobes free at tips, about 2 mm long, with margins incurved about middle and tips thickened and incurved.

Brachystelma occidentale, one of the minutest species in the genus, may never again be seen in the wild. The concept of this *western* species is based on a single specimen collected by R.Schlechter in a stony place on the mountain side near Smitswinkel Bay on the Cape Peninsula. The question is whether to record it as a seriously endangered species or one probably extinct. A few

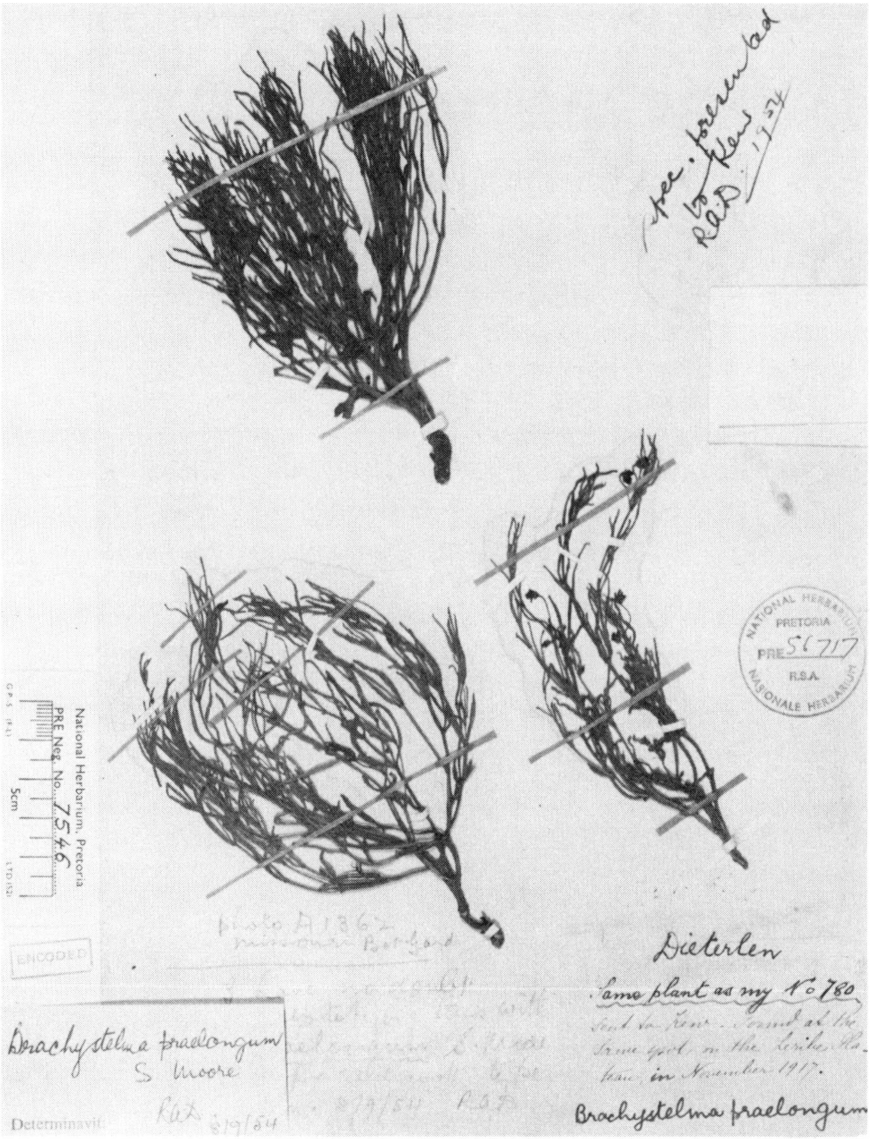

Figure 27. *Brachystelma praelongum:* sheet in herbarium BRI (specimen collected by *Dieterlen* on Leribe Plateau, Lesotho, and compared with type)

years back botanists at the Cape made every effort to rediscover the species but without success. Harry Hall reported at the time that there was no cause for optimism in the face of alien plant invasion and the advance of modern housing estates.

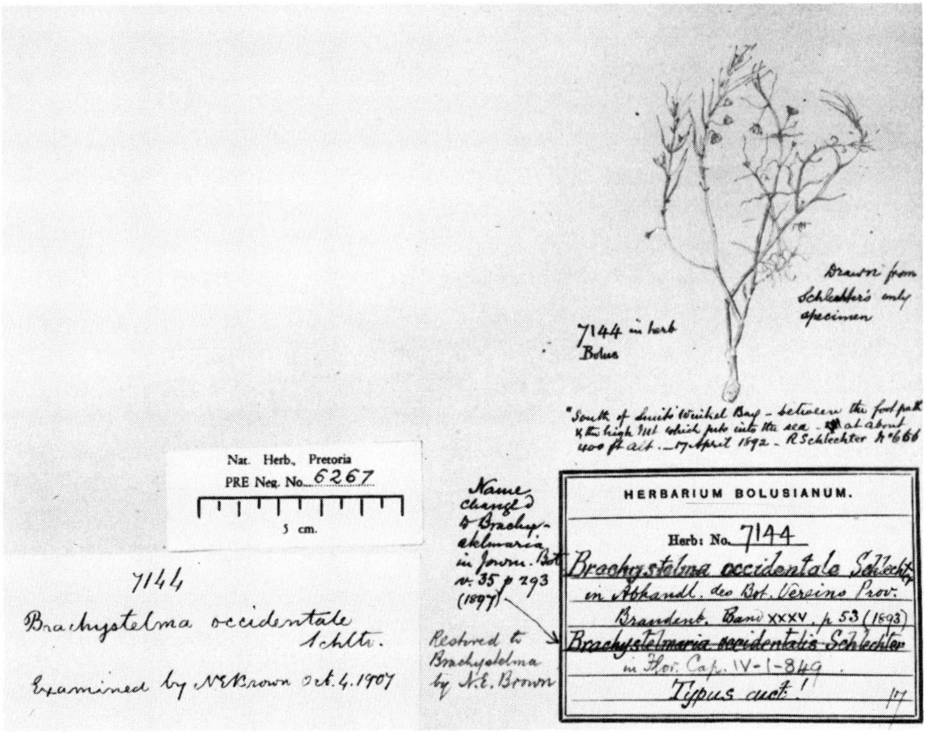

Figure 28. *Brachystelma occidentale:* this is a copy of a drawing of the only specimen collected of this species in the history of the Cape. The specimen was collected by *R.Schlechter* no.666, on 17 April 1892, south of Smits Winkel Bay, near Simonstown, on a rocky ridge at about 100 m altitude and possibly never to be seen alive again.

27. **Brachystelma nanum** *(Schltr.) N.E.Br.* in Fl.Cap. 4,1:848 (1908); R.A. Dyer in Flora S.Afr. 27,4:21 (1980) [Fig.29]

Perennial dwarf herb with tuber. *Tuber* subglobose up to about 40 mm diameter. *Stems* 2-several produced annually, 50-80 mm tall, slender. *Leaves* narrow up to 20 mm long, tapering into a short stalk. *Flowers* 1-3, lateral at the nodes. *Corolla* up to 4,5 mm long, mainly pale pink to whitish with green tips; tube minute ±1 mm long, cup-like round corona; lobes 3,5 mm long, slightly thickened and incurved at tips.

Brachystelma nanum, so named because of its dwarf habit, reveals its beauty only when one is able to study the flowers under an enlarging lens and see the perfect mechanism of the coronal structure. Both Schlechter and N.E.Brown drew attention to the apparently close relationship between *B.nanum* from northern OFS and Transvaal and *B.occidentale* from the Cape Peninsula. It seems certain that structural differences would become apparent were it possible to

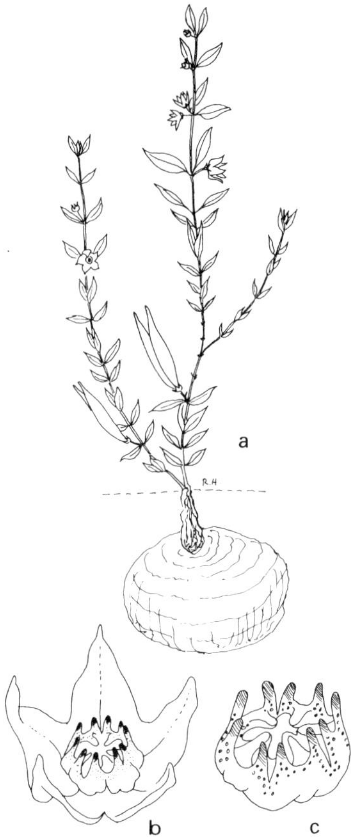

Figure 29. *Brachystelma nanum:* a. plant, x ½; b. open flower with corona in centre, x 5; c. corona with outer lobes bilobuled (5 x 2), inner lobes incumbent on the staminal column, x 8 (after *Liebenberg* in BRI)

study live specimens side by side. In present circumstances one has to rely on a colour difference in the corolla, yellowish spotted with wine red in *B.occidentale* and pinkish with dark green tips in *B.nanum. B.nanum* unlike *B.occidentale* from far south on the Cape Peninsula, is quite widespread in the Transvaal and sometimes locally common round Johannesburg.

28. **Brachystelma micranthum** *E.Mey.*, Comm.fas.2:196 (1838) (Stafleu); R.A. Dyer in Flora S.Afr. 27,4:21 (1980)

Perennial dwarf herb with tuber. *Tuber* probably about 40 mm diameter and flattened from above. *Stems* few produced annually, 50-70 mm tall, minutely hairy. *Leaves* narrow, 10-12 mm long, up to 6 mm broad. *Flowers* 1-2 at nodes. *Corolla* very minute, 2 mm long; lobes slightly longer than the calyx.

Brachystelma micranthum, one of the several very 'small flower' species in the genus, presents a real problem in the identification because it is known

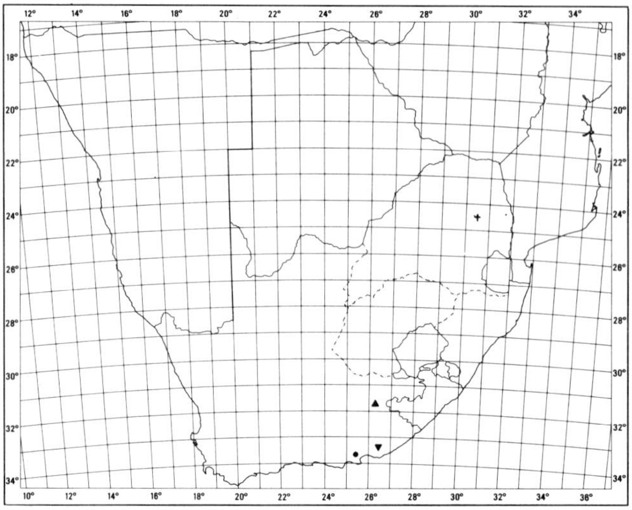

▲ *Brachystelma micranthum* (28)
▼ *Brachystelma delicatum* (29)
● *Brachystelma schonlandianum* (30)
+ *Brachystelma parvulum* (31)

only from the type specimen which is now fragmentary. The locality of origin was in the Queenstown district of the eastern Cape in the vicinity of the old mission station at Shiloh. The habitat must have altered appreciably in the past 150 years and a rediscovery at this stage would be an event to celebrate.

29. **Brachystelma delicatum** *R.A.Dyer* in Bothalia 12,1:53 (1976); Flora S.Afr. 27,4:21 (1980) [Plate 8, Fig.30]

Perennial dwarf herb with tuber. *Tuber* subglobose, 20-40 mm diameter. *Stems* 1-3, produced annually, erect, up to about 60 mm tall, slightly zig-zag, unbranched or sparingly branched from near base, reddish. *Leaves* narrow, up to about 25 mm long, 2-7 mm broad, somewhat fleshy. *Flowers* 4-7 together, slightly lateral at the nodes, opening more or less together. *Corolla* 3-4 mm long, with a small bell-shaped tube; tube 1-1,25 mm deep, 1,25 mm diameter at mouth; lobes 2-3 mm long, 1-1,5 mm broad at base, white with green or brown slightly thickened tips.

Brachystelma delicatum is truly dainty and charming and as often happens its discovery came by chance and good fortune. In the only two records of the species the dormant tubers were exposed when the collectors were seeking succulent plants near rocky outcrops in grassveld. First Dave Hardy found tubers in September 1965 in Trappes Valley near Grahamstown in the eastern Province and these flowered at BRI in December of the same year and on subsequent occasions. The second discovery was by Roy and Estelle Bayliss in August 1968 not far distant in Coombes Valley. These tubers flowered in

Figure 30. *Brachystelma delicatum:* plant with tuber and clusters of flowers at the nodes, cult. BRI, x 1 (*Hardy* 2198, near Grahamstown, eastern Cape)

Grahamstown also in December of the same year when one small tuber produced 28 flowers lasting about 12 days. Roy Bayliss, who took beautiful photographs, remarked that the individual flowers were smaller than a match head.

30. **Brachystelma schonlandianum** *Schltr.* in Bot.Jb. 18 Beibl. 45,35 (1894); R.A.Dyer in Flora S.Afr. 27,4:22 (1980)

Perennial dwarf herb presumably with small tuber. *Stem* produced annually up to about 50 mm tall, branched. *Leaves* slender, up to 10 mm long, tapering into a short stalk. *Flowers* 2-4 together at the nodes. *Corolla* about 5 mm; tube 3 mm long, bell-shaped 2 mm diameter at mouth; lobes 2 mm long.

Brachystelma schonlandianum was one of the minor tragedies of the 1939-1945 war; the type specimen of this diminutive species was destroyed with many others in the Berlin-Dahlem herbarium. It had been collected by Schlechter personally on dry hills near Uitenhage in the eastern Cape and named after his personal friend Selmar Schonland, later professor of Botany at Rhodes University, Grahamstown. Unless a plant is found in the type locality agreeing with the description, it may eventually have to be assumed to be an extinct species. It must have had some likeness to *Brachystelma delicatum* but according to the description differs appreciably in the ratio of the lengths of corolla-lobes to corolla-tube.

31. **Brachystelma parvulum** *R.A.Dyer* in Bothalia 12,3:447 (1978); Flora S. Afr. 27,4:22 (1980) [Fig.31]

Perennial dwarf herb with tuber. *Tuber* globose, about 20 mm diameter. *Stem* single annually, unbranched or perhaps sometimes once-branched, up to 50 mm tall. *Leaves* slender, with short stalk, up to about 25 mm long and 4 mm broad. *Flowers* 1-2, lateral at nodes developed successively, with slender stalks about 12 mm long. *Corolla* 4-5 mm long, 6-9 mm diameter, divided to about half-way; tube small bell-shaped, about 2 mm deep, whitish within; lobes triangular, about 2,25 mm long, with margins slightly recurved towards tip, dark maroon.

Brachystelma parvulum is another species named in reference to its small size. It must be among a very few species which owe their discovery to the assistance in the field of a helicopter. As a special favour in February 1977 Dave Hardy of BRI was landed during a military exercise on the summit of the Transvaal Drakensberg above the Strydom tunnel. Dave Hardy's trained eyes recognised at his feet in short grass a small flower which had to be a *Brachystelma.* He collected the tuber which flourished at BRI and has flowered during each subsequent year. The illustration gives a good idea of the delicate nature of the species and the face view in 'c' shows the unusual corona at the base of the corolla lobe.

32. **Brachystelma minor** *E.A.Bruce* in Flower.Pl.Afr. 28:t.1096a (1951); R.A. Dyer in Flora S.Afr. 27,4:22 (1980) [Fig.32]

Perennial dwarf herb with tuber. *Tuber* up to about 50 mm diameter, compressed about 20 mm thick. *Stem* single annually, up to 30 mm tall or spread-

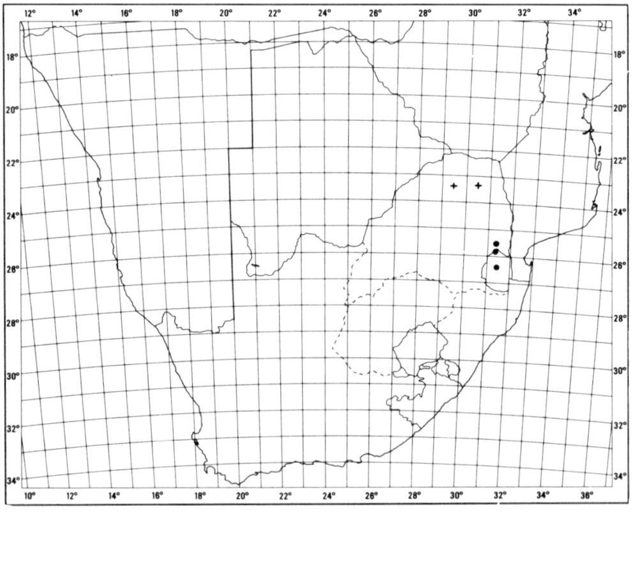

+ *Brachystelma minor* (32)
● *Brachystelma swazicum* (33)

Figure 31. *Brachystelma parvulum:* a. type plant in cultivation at BRI from summit of Transvaal Drakensberg in the Pilgrim's Rest district, x 1; b. side view of flower showing corolla tube, x 5; face view of flower revealing corona, x 5 (*Hardy* 4090, photo: Adela Romanowski)

ing up to 100 mm in shade. *Leaves* comparatively dense with stalk 8-16 mm long; blade up to about 20 mm long and 10 mm broad, purple-tinged. *Flowers* 2-3, lateral at nodes developing successively on short stalks and often somewhat hidden by leaves. *Corolla* about 7,5 mm long, divided about one-third

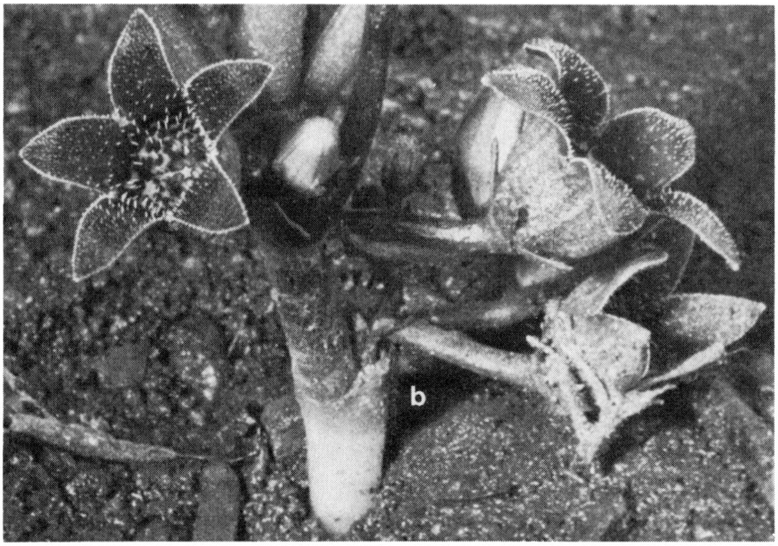

Figure 32. *Brachystelma minor:* a. the type plant in cultivation at BRI 1949, showing the characteristic habit of flowers on considerably shorter stalks than the leaves by which the flowers are often partially hidden, x 1; b. showing obtuse bud in centre and hairiness of corolla, x 5 (*Murray* in BRI)

its length; tube bell-shaped, 4-5 mm long, 5 mm diameter, greenish-cream outside with red venation, maroon or purple within with greenish-cream streaks; lobes triangular, slightly spreading with short white bristle-like hairs within and densely ciliate.

Brachystelma minor is a small species in which the attractive little flowers are liable to be hidden by the longer leaves. A similar habit is seen in *Brachystelma oianthum* but in this species the flowers are appreciably larger and project from the basal leaves. Eileen Bruce recounts in Flowering Plants of Africa that *B.minor* was discovered in October 1947 by Joan Murray, growing between dolomitic rocks near The Downs of northern Transvaal. The distribution range has been extended by collections on the Wolkberg and near Graskop of the eastern Transvaal where it may be locally common, tolerating both open and shady conditions.

33. **Brachystelma swazicum** *R.A.Dyer* in Bothalia 12,1:56 (1976); Flora S.Afr. 27,4:23 (1980) [Plate 9, Fig.33]

Perennial prostrate herb with tuber. *Tuber* up to about 50 mm diameter. *Stems* 2-3 produced annually, prostrate, rarely branched, up to about 250 mm long, with internodes 15-20 mm long,red. *Leaves* up to about 20 mm long and 20 mm broad, sometimes slightly heart-shaped at base. *Flowers* 2-3(4) together on a rudimentary stalk, developing successively. *Corolla* about 4,5 mm long, shortly tubular, dark purple; tube broadly bell-shaped or saucer-shaped, 1,5-2 mm long, 3 mm wide; lobes triangular, 3 mm long.

Brachystelma swazicum, a charming little plant, has a prostrate habit. At first this led to some confusion between this species and *Brachystelma pulchellum* from Natal. It is, however, readily distinguished from this by the shortly

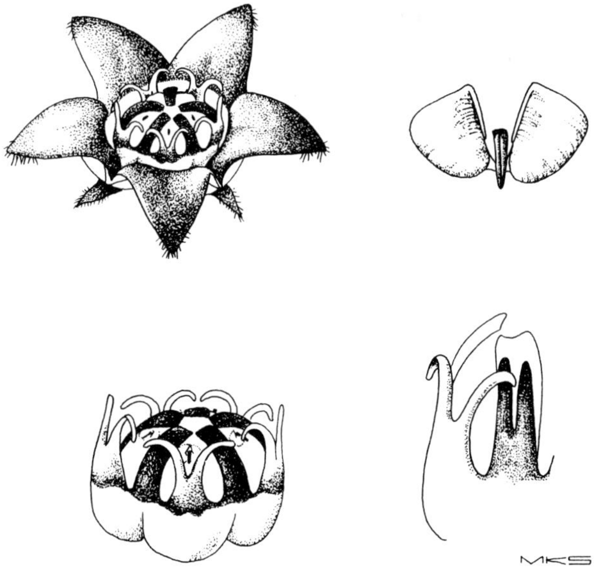

Figure 33. *Brachystelma swazicum:* a, flower face view with corona in tube, x 5; b, corona showing deeply divided outer lobes with spreading lobules, and incumbent inner lobes, x 10; c, one outer and one abnormally divided inner lobe, x 30; d, pollinia with short connectives to wingless carrier, x 50 (*Omerod & Lavranos,* photo Adela Romanowski)

pedunculate inflorescence with a few long hairs round the attachment and more certainly by the falcately curved lobules of the outer corona lobes. It has been found to be fairly frequent on the highest mountain ridges of Swaziland and extends over the common border with eastern Transvaal in the Barberton district.

34. **Brachystelma tabularium** *R.A.Dyer* in Bothalia 12,4:629 (1979); in Flora S.Afr. 27,4:34 (1980) [Fig.34]

Perennial herb with tuber (presumed). *Stem* produced annually, 2-3 branched from near base, erect, about 200 mm tall with internodes 25-70 mm long, shortly hairy. *Leaves* shortly stalked; blade about 10 mm long and 5 mm broad, somewhat undulate on margin. *Flowers* 1-2 or more lateral at the nodes, opening in succession. *Corolla* cylindric in bud, truncate with the tips of the lobes inflexed, about 12 mm long, divided almost to the base; lobes linear with free inflexed tips, about 12 mm long and 0,5 mm broad in the middle and base 1,25 mm.

Brachystelma tabularium is based on a specimen collected by Florence Paterson at Redhouse near Port Elizabeth in the eastern Cape in 1913. She sent the tuber to the Bolus Herbarium at Cape Town University where it flowered in 1915. A botanical drawing was then made by A.C.Dawson, an assistant at the herbarium and friend of the Bolus family but possibly because of the death of Harry Bolus about that time no description was made and the illustration found its way into the herbarium archives. It was that event which suggested the specific name. The drawing was retrieved from oblivion by Prof. E.A.Schelpe in 1978 and kindly sent on loan to BRI where it was incorporated into the revision of the genus for the Flora of Southern Africa. No matching specimen has been recorded in the interim of over 60 years.

This long period during which no further record has been made recalls other examples from the same area. There has been no rediscovery of *B.schonlandianum* in 85 years, but, on the other hand, very recently Roy Bayliss rediscovered *B.comptum* after a lapse of time of nearly 150 years and Estelle Bayliss did the same for *B.campanulatum* after a similar sterile period.

The cylindric truncate bud of *B. tabularium* with free inflexed tips of the corolla lobes is unique in the genus.

Plate 7. *Brachystelma pulchellum* (species 19); a. plant with tuber slightly exposed, usually submerged in nature, flowers 2 at a node developed in succession, x 1; b. flower face view with traces of circular marking, calyx segments exposed well beyond sinuses, x 4; c. corona with no basal tube, inner lobes incumbent, x 10; d. pollinia very shortly attached to winged carrier, x 60 (*O.Hilliard & B.L.Burtt* 8590, near Krantzkloof, Natal)

Plate 8. *Brachystelma delicatum* (species 29); a. plant grown at BRI and flowered in 1976, x 1; b. flower with white corolla marked with dark maroon-brown to lime-green at tips of libes, x 8; c. corona with shortly spreading emarginate outer lobes and incumbent inner lobes, x 15; d. pollinia very shortly attached to winged carrier, x 60 (*D.Hardy* 2198, south of Grahamstown, eastern Cape)

c
d
b
a
Marguerite Scott

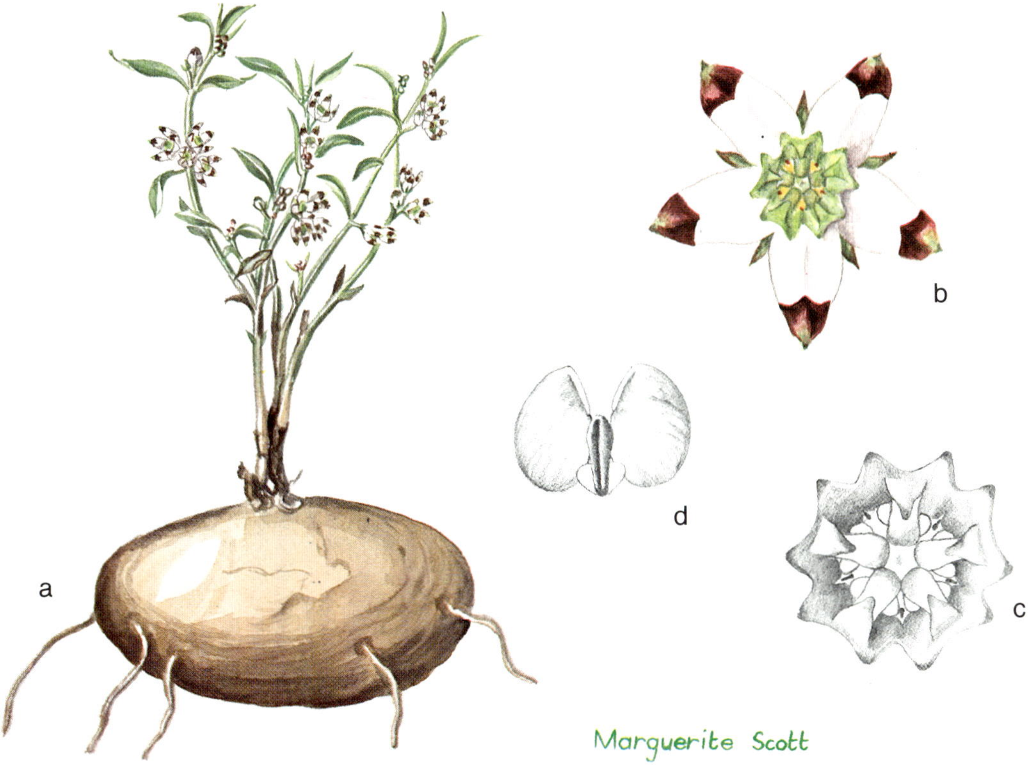
a
b
d
c
Marguerite Scott

b
c
d
a

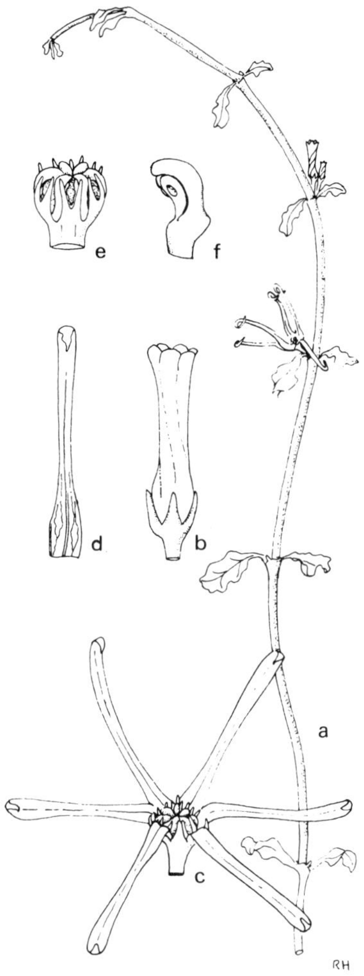

Figure 34. *Brachystelma tabularium:* a. branch, x ½; b. bud, x 2; c. flower opened out, x 2; d. corolla segment with infolded tip, x 2; e. corona, x 4; f. anther, x 10 (from a drawing by A.C.Dawson in Bolus Herbarium archives, of specimen coll. *Florence Peterson* in 1913 from Redhouse near Port Elizabeth, eastern Cape)

35. **Brachystelma huttonii** *(Harv.) N.E.Br.* in F.C. 4,1:845 (1908); R.A.Dyer in Flora S.Afr. 27,4:23 (1980) [Plate 10, Fig.35]

Perennial dwarf herb with tuber. *Tuber* about 50 mm diameter, depressed. *Stem* single produced annually, sparingly to freely branched, 80-120 mm tall, slightly zig-zag on flowering part, reddish, minutely hairy. *Leaves* slender, 15-25 mm long, 1-3 mm broad, tapering to base. *Flowers* 1-10 together, lateral at both sides of the nodes and terminal. *Corolla* yellow to maroon, shortly

---

Plate 9. *Brachystelma swazicum* (species 33); a. plant in cultivation showing prostrate habit of stems with leaves on upright petioles (stalks) and shortly stalked flowers, 1-2(3) opening in succession at hairy nodes, x 1; b. flower, x 5; c. corona with deeply bilobulate outer lobes and incumbent-erect maroon inner lobes, x 10; d. pollinia shortly attached to wingless carrier, x 50 (*J.Lavranos* in PRE 50301, Swaziland)

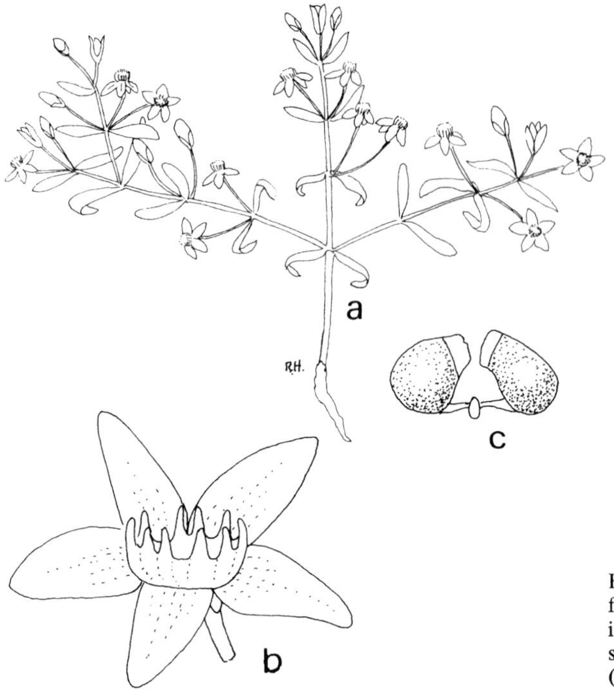

Figure 35. *Brachystelma huttonii:* a. flowering stems, x ½; b. flower exposing corona, x 3; c. pollinia attached to small carrier without wings, x ±15 (after *Harvey,*Thes. 2:9, t.114)

united at base; tube about 1 mm deep round base of corona; lobes free, about 3 mm long, with reflexed margins.

Brachystelma huttonii is named after Henry Hutton who pioneered some of the more important roads of the eastern Cape from 1850 to 1860. It was in the karroid scrub of the Fish River Valley not many miles north from Grahamstown that he discovered *Brachystelma huttonii,* when surveying the road to Fort Beaufort and over the Katberg to Queenstown. Little more was known of the species until fairly recent times when re-collected in the same environment, first by Harry Hall and later by Roy Bayliss. Harry Hall recorded a wide range of colouration, from yellow to maroon, of flowers on different plants in the arid scrub at Pluto's Vale where the specimens were locally abundant.

36. **Brachystelma arnotii** *Bak.* in Refug.bot. 1:t.9 (1869), as *arnottii;* R.A. Dyer in Flora S.Afr. 27,4:23 (1980) [Fig.36, 36A]

Perennial dwarf herb with tuber. *Tuber* 50-60 mm diameter, somewhat depressed. *Stems* 1-few produced annually, erect, simple or sparingly branched, 70-200 mm tall, minutely decurved hairy. *Leaves* broad to narrow up to 40 mm long, tapering into short stalk, margins usually folded upwards and

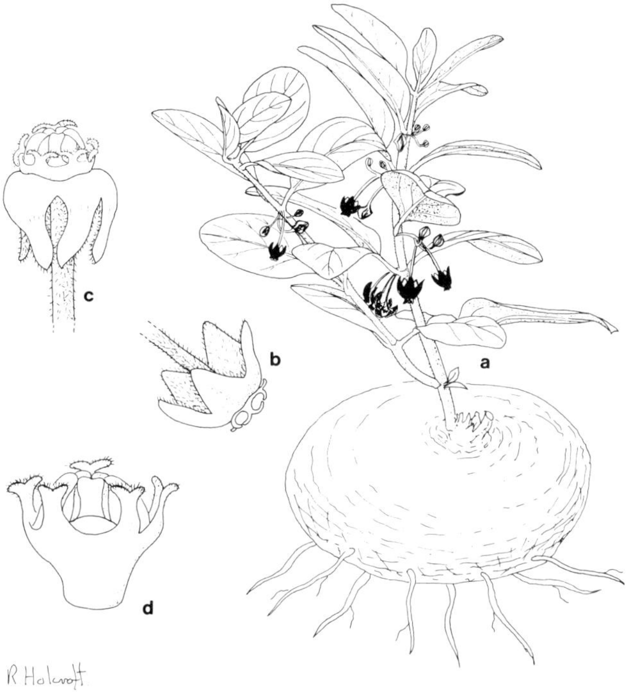

Figure 36. *Brachystelma arnotii:* a. plant, x ½; b. flower showing reflexed corolla lobes and exposed corona, x 3; c. different view of flower, x 3; d. corona with bifid outer lobes and cushion-like inner lobes pressed against base of anthers, x 10 (after Refugium Bot. 1869 and *J. Verhagen* in BRI)

somewhat undulate. *Flowers* 2-6 together lateral at the nodes, opening successively or sometimes 2 together, on slender stalks. *Corolla* 6-10 mm diameter, deeply divided, without tube; lobes 2-3 mm long, spreading-reflexed, purple-brown with green tubercle-like thickening at tip.

Brachystelma arnotii is named after David Arnot whose name is rendered variously in literature as *arnottii* and *arnoldi.* In the middle 1800's he was in residence in Colesberg in the north eastern Cape from whence he posted specimens to Kew. These were not adequately labelled with locality records and some were obviously not from the environs of Colesberg. Arnot was at the time the representative of the Griquas, west of Kimberley, and had widespread contacts along that route. Thus it can be assumed that some of his specimens came from various stations other than Colesberg. It is of some importance in the case of *Brachystelma arnotii* because the species has been traced from northern SW Africa (Namibia), through Botswana into Northern Cape and possibly into the sandveld of western Transvaal. There is no specimen authentically from Colesberg. In SW Africa specimens were given the name *B.grossartii* by Dinter but this was considerably later than *B.arnotii* of Baker, of which it is regarded as a form. In describing *B.grossartii,* Dinter remarked that the variation in the length and width of the leaves is so great that one may easily think one is confronted by two completely different plants (species) when dealing with

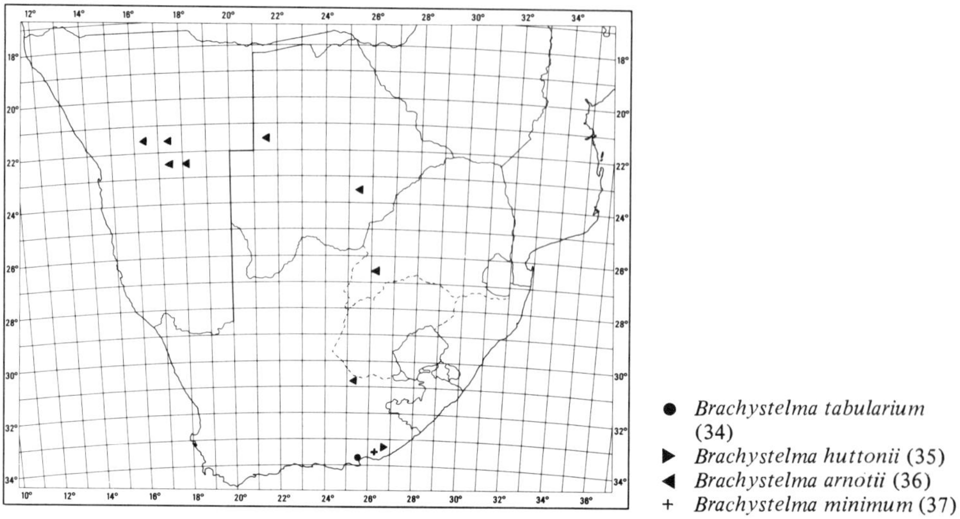

Figure 36A. *Brachystelma arnotii:* flowering peduncle, one old flower, side view of two open flowers and two buds, x 5 (*J. Verhagen* 12, photo Adele Romanowski)

extreme forms. Dinter added that the same phenomenon is met with in the case of *Brachystelma dinteri, B.(Siphonostelma) stenophyllum, Ceropegia pygmaeum* and several other related species in SW Africa.

Brachystelma arnotii is notable for its strongly reflexed corolla lobes and the exposed corona, the outer lobes of which are deeply divided with spreading lobules.

Figure 37. *Brachystelma minimum:* a. enlarged habit, x 2; b. flower with corolla divided ⅔ to base, lobes reflexed to pedicel, corona very delicate and transparent except for margin, x 4; c. drawing showing densely ciliate corolla lobes recurved to pedicel, x 20 (*M.B. Bayer* 821, cult. BRI, photo: Adela Romanowski)

37. **Brachystelma minimum** *R.A.Dyer* in Bothalia 12,3:447 (1978); in Flora S.Afr. 27,4:24 (1980) [Fig.37]

Perennial dwarf herb with tuber. *Tuber* about 35 mm diameter, depressed. *Stems* 2-3, produced annually, very slender, up to about 50 mm tall. *Leaves* slender, 10-12 mm long, about 1,25 mm broad. *Flowers* solitary lateral at the nodes on very slender stalks about 12 mm long. *Corolla* about 3 mm long, 4,5 mm diameter, divided half to two-thirds to base; tube at first 1,25 mm deep, 2 mm diameter, white round base of corona; lobes triangular, about 1,5(2) mm long, the tip gradually recurved to the stalk and corolla becoming button-shaped, purple-maroon, with long white cilia on margin and the tubular corona prominently exposed on the centre of the corolla.

Brachystelma minimum, by its name, claims to be one of the smallest species in the genus. One does not wish to dramatise too sharply the discovery of a species new to science, but in the present case it is safe to surmise that without the collector's acute perception the diminutive plants would have gone unnoticed for a further long period. The discovery was made by Bruce Bayer in the eastern Cape south of Grahamstown overlooking the Bushman's river, only a matter of a few days before the equally interesting discovery of *B.parvulum* was made by Dave Hardy in the northern Transvaal. Bruce Bayer was at first unimpressed by his find but the tubers flowered again at the Karoo Botanic Garden, Worcester, when he gave them closer scrutiny and drew them to my attention.

Not only is the corolla of *B.minimum* very small but the fringed lobes recurve to the flowering stalk into an even smaller button-shaped whole. In the recurving of the corolla lobes it shows a resemblance to *B.arnotii* and *B.huttonii,* the latter found under the protection of karroid scrub in the Fish River Valley north of Grahamstown. The environs of Grahamstown support several other rare species of the genus.

38. **Brachystelma chlorozonum** *E.A.Bruce* in Hook.Icon.Pl. t.3370 (1938); R.A.Dyer in Flora S.Afr. 27,4:24 (1980) [Fig.38, 38a]

Perennial dwarf herb with tuber. *Tuber* up to about 70 mm diameter, somewhat depressed. *Stem* 1, produced annually, sometimes sparingly branched, up to about 100 mm high (200 mm in cultivation). *Leaves* shortly stalked, 20-45 mm long, 15-25 mm broad. *Flowers* solitary, lateral at nodes, with the short stalk usually somewhat decurved. *Corolla* about 15 mm diameter, divided to about the middle, yellowish-green outside; tube broadly bell- or basin-shaped, 4-6 mm deep, 8 mm diameter, with greenish to almost black ring-markings within, minutely green-spotted round corona; lobes triangular, 4,5-5 mm long, spreading-recurved, with white or purplish hairs on inner surface.

Brachystelma chlorozonum is surrounded by uncertainty as to its exact locality of origin. The original tuber was sent to the Royal Botanic Gardens, Kew early in the century from Barberton, eastern Transvaal, by George Thorncroft. It has not been recorded from Barberton since but it has been collected by Dave Hardy on the eastern Transvaal-Mozambique border. Other records

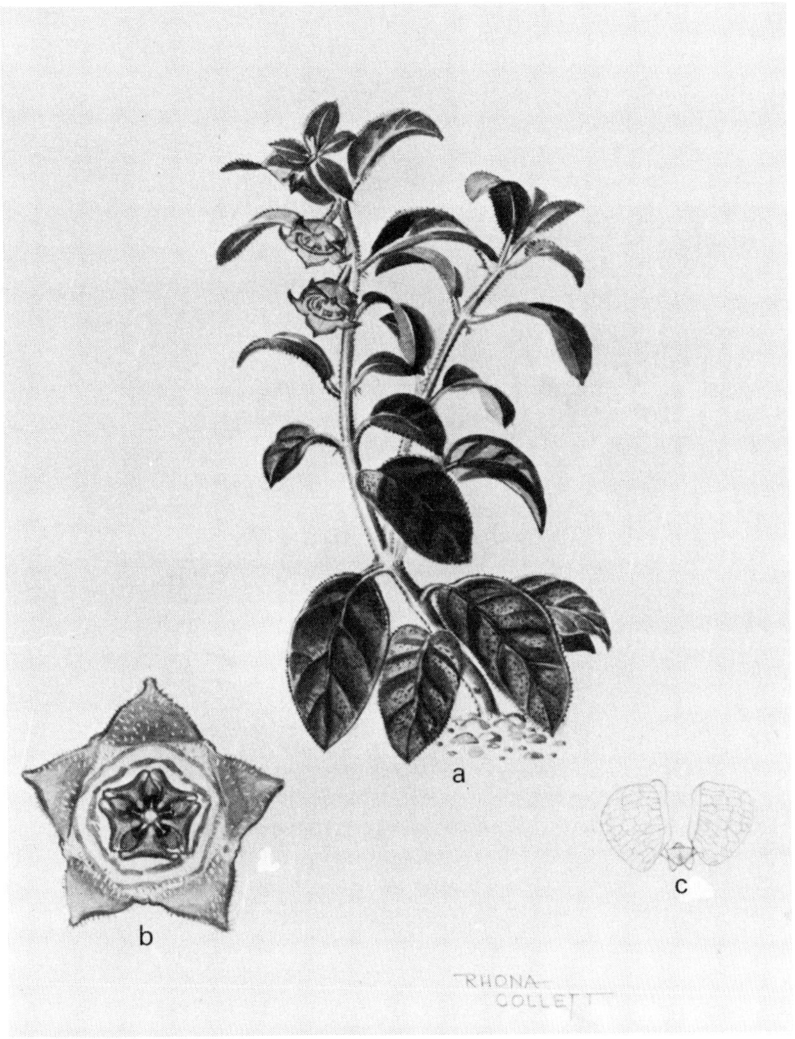

Figure 38. *Brachystelma chlorozonum:* a. plant from northern Zululand, drooping flower and bud together (sometimes only one flower), x ¾; b. flower face view with corona at base of bell-shaped corolla tube, two circular bands of colour round upper part of tube, x 3; c. pollinia attached by very short caudicles (connectives) to small winged carrier, x 15 (*Vahrmeijer & Hardy* 637, cult. BRI)

have been verified from Zululand where it is claimed to be locally common. From there Andre Liebenberg noted that the banded colouration of flowers, on which the species was named, varies from 1-5 dark rings of colour within the corolla-tube and other parts were distinctly reddish. It seems fairly consistent that the somewhat bell-shaped flowers are on stalks which curve downwards as if to express modesty.

Figure 38a. *Brachystelma chlorozonum:* face view of corolla showing corona at base with slight variation from Fig.38 and only one colour band round corolla tube, x 6 (*A.Liebenberg* 464, cult BRI)

39. **Brachystelma ngomense** *R.A.Dyer* in Bothalia 12,2:255 (1977); in Flora S.Afr. 27,4: (1980) [Plate 11, Fig.39]

Perennial prostrate herb with tuber. *Tuber* subglobose, 20-30 mm diameter. *Stems* 1-3, produced annually, prostrate, sparingly branched, up to about 100 mm long (200 mm in cultivation). *Leaves* shortly stalked up to about 15 mm long and 10 mm broad, narrowed from slightly below middle. *Flowers* 1-2, slightly lateral at nodes on stalks about 10 mm long. *Corolla* slightly bulged above base in bud, 10 mm long, divided to about half-way; tube cylindric-bell-shaped, about 5 mm long and 5 mm diameter, with flattish base, slightly in-fluted from the base of the lobes, making the mouth obtusely 5-angled, white

Figure 39. *Brachystelma ngomense:* a. showing diffuse subprostrate branching from tuber, x 1; b. flower showing light-coloured corona at base of corolla surrounding the dark staminal column, x 4 (*O. Hilliard & Burtt* 8441, cult. BRI, photo: Adela Romanowski)

within and finely purple-mottled; lobes triangular, about 5 mm long, spreading recurved, dark red or maroon, with white tip.

Brachystelma ngomense was discovered by Olive Hilliard and Bill Burtt in northern Natal near Ngome. At the time the collectors were, among other interests, plotting distribution records of *Brachystelma pulchellum* (sp.19). *B. ngomense* has been found only in a small area near Ngome on the edge of sheets of exposed rock. It is a delightful addition to the wealth of corolla forms within the genus and has no close affinity to any known species. Tubers were grown successfully at BRI where good illustrations were obtained.

40. **Brachystelma incanum** *R.A.Dyer* in Bothalia 12,1:54 (1976); in Flora S. Afr. 27,4:25 (1980) [Fig.40]

Perennial spreading decumbent herb with tuber. *Tuber* about 50 mm diameter. *Stem* single (2) produced annually, branched from base; branches spreading-decumbent up to about 100 mm long, hairy. *Leaves* shortly stalked, 15-30 mm long, 10-20 mm broad, hairy on both surfaces or only on lower, sometimes folded upwards. *Flowers* 1-4, lateral at the nodes, developing in succession. *Corolla* star-shaped from above, up to 10 mm diameter, two-thirds-divided; tube short containing base of corona and sometimes spreading to base of lobes,

Figure 40. *Brachystelma incanum:* a. typical form (dried) from south of Lichtenburg, Transvaal, showing dense ring of long hairs round basin-shaped corona with staminal column in centre, x 3 (*Acocks* 12476); b. habit, less silvery hairy than typical form, x 1½; c and d. two views of inflorescences with typical circle of hairs round corona, x 2½ (*A.E.van Wyk* in BRI 57637, Wolmaransstad, photo: Adela Romanowski)

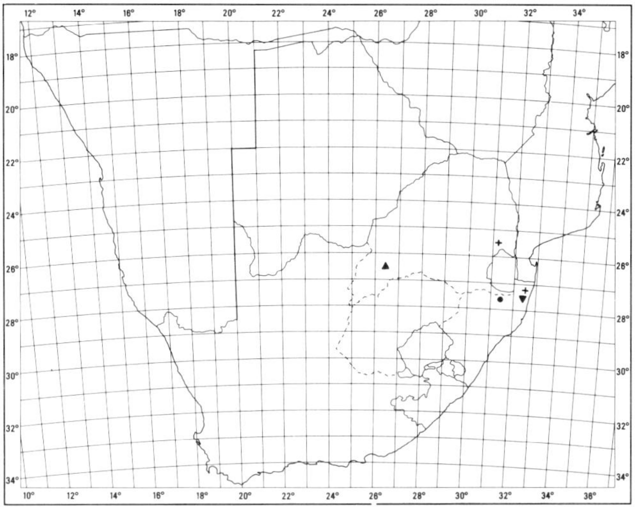

+ *Brachystelma chlorozonum* (38)
● *Brachystelma ngomense* (39)
▲ *Brachystelma incanum* (40)
▼ *Brachystelma vahrmeyeri* (41)

hairs light violet, drying whitish; lobes triangular, dark violet-black, with wrinkled surface and smooth greenish-yellow tips.

Brachystelma incanum was established on a single specimen collected by John Acocks in 1946 from the western Transvaal, 22 km SW of Lichtenburg. In its dried state it was decidedly silvery hairy, therefore the name, but later collections proved the chosen specific name to be somewhat misleading. In January 1979 a specimen collected by A.E.van Wijk from near Wolmaransstad was brought to the BRI for identification. It was generally less silvery hairy than the Acocks specimen except for the hairs round the corona: the similarities became more evident as the drying process continued and I have little doubt that they represent only one species, variable as is usual. After about one year in cultivation one tuber sent up a short erect stem with normal flowers. Presumably this could also happen in nature under unusual environmental conditions.

41. **Brachystelma vahrmeijeri** *R.A.Dyer* in Bothalia 10:278 (1971); in Flora S. Afr. 27,4:26 (1980) [Fig.41]

Perennial tufted herb with tuber. *Tuber* up to about 50 mm diameter, slightly compressed, red-fleshed. *Stems* several annually, up to 100 mm tall, rarely rebranched above. *Leaves* slender, up to 30 mm long, 7 mm broad with little or no stalk. *Flowers* 2-3 together, lateral at nodes, the oldest uppermost, opening in succession on stalks 5-10 mm long. *Corolla* funnel-shaped, 8 mm long, 4 mm diameter at mouth, yellow-green or cream, rarely white or maroon; tube 3 mm long; lobes more or less triangular, 5 mm long, slightly spreading.

Figure 41. *Brachystelma vahrmeijeri:* type plant from northern Zululand near coast, x 1 (*H. Vahrmeijer* 1050)

Brachystelma vahrmeijeri, unlike most other species which have been collected once only, was recorded as frequent in its habitat in the coastal belt of northern Zululand. Hans Vahrmeijer then of BRI staff also collected specimens of *Brachystelma tenue* (sp.52), another previously undescribed species, under similar conditions. One feels relatively confident that if need be one could return to the area and have little difficulty in locating a supply of tubers. In the case of species based on a single specimen, it usually implies that plants are rare and it may be very many years before they are rediscovered, as in the case of *B.campanulatum* no.13.

42. **Brachystelma cupulatum** *R.A.Dyer* in Bothalia 10:375 (1971); Flora S.Afr. 27,4:26 (1980) [Fig.42]

Perennial dwarf herb with tuber. *Tuber* up to about 100 mm diameter, subglobose or sometimes somewhat depressed. *Stem* single, produced annually,

Figure 42. *Brachystelma cupulatum:* a. plant in flower with tuber, x 1; b. buds and face view of flower with cup-shaped corona at the base of which are the pollinia, x 5 (*J. Vahrmeijer* 2345, cult. BRI, photo Adela Romanowski)

sometimes once-branched from near base, sparsely branched above, up to 150 mm tall. *Leaves* 30-80 mm long, 4-15 mm broad, contracted into a short stalk at the base. *Flowers* 1-few lateral at the nodes, developing successively on short stalks. *Corolla* 6-9 mm long, green; tube 1,5-2,5(3) mm long; lobes about 4,5-6 mm long, with slightly recurved margins.

Brachystelma cupulatum was given that name because of the cup-like shape

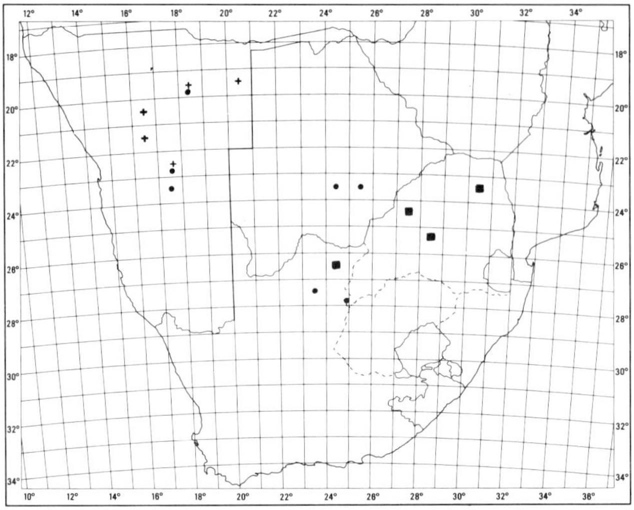

- ● *Brachystelma cupulatum* (42)
- \+ *Brachystelma dinteri* (43)
- ■ *Brachystelma brevipedicellatum* (43A)

of the corona at the base of the corolla. This feature is absent in *B.blepharanthera,* with which it seems to have been confused when first collected by Dinter in SW Africa early in this century. The fact that the flowers are dark green allows them to pass almost unnoticed. On the other hand the species is of importance in the life of Bushmen because of the edibility of the tubers which occur in some quantity from the sandveld of northern SW Africa into the Kalahari sands of Botswana and northern Cape.

43. **Brachystelma dinteri** *Schltr.* in Bot.Jb. 51:144 (1913); Dinter Neue Pfl. S.W.Afr. 15, fig.7 (1914); R.A.Dyer in Flora S.Afr. 27,4:26 (1980), excluding *B.brevipedicellatum* Turrill and *B.ringens* E.A.Bruce [Fig.43]

Perennial dwarf herb with tuber. *Tuber* 30-50(120) mm diameter, somewhat depressed from above. *Stem* erect, about 3 mm diameter, simple or sparingly branched, 150-200(400) mm tall, with internodes 15-20 mm long, shortly hispid, leafy. *Leaves* spreading-erect, elliptic or occasionally lanceolate-elliptic, ±40(50) mm long, ±16(10) mm broad, obtuse or subacute, shortly stalked or petiolate, shortly hairy on both sides. *Flowers* lateral at nodes, fasciculate, shortly pedicellate, (few)9-12 together, developing in succession but sometimes 2-3 open together. *Sepals* lanceolate, densely puberulous. *Corolla* subrotate, ±10 mm diameter, divided to slightly more than halfway, with no appreciable tube (Dinter); lobes ovate-triangular, acute or subacute, densely puberulous on outer surface, green; glabrous on inner surface and somewhat rugose, bright green with brown concentric blotches or dark green with larger

Figure 43. *Brachystelma dinteri:* preserved flower showing slight corrugations on corolla lobes and broad corona united at base into short tube, 5-lobed above, the lobes subquadrate, emarginate, ascending, incurved, considerably exceeding the height of the staminal column and laterally confluent at base with inner corona lobes; inner lobes more or less oblong, pressed to base of anthers but not extended over their backs, x 5 (*P.Bruyns* 1986, photo: Adela Romanowski, for comparison with *B.brevipedicellatum*)

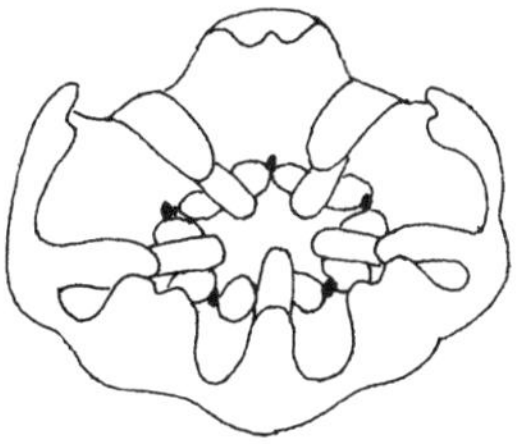

blotches (Dinter). *Corona* saucer or broadly cup-shaped in outline, united in lower half, 2,5-3 mm diameter; outer lobes subquadrate, emarginate or 2-toothed overtopping staminal column, laterally confluent with base of inner lobes; inner lobes incumbent on base of anthers, not extending over staminal column. *Pollinia* obliquely quadrate-ellipsoid, attached at base to small carrier by short caudicles.

Brachystelma dinteri is named after the famous German collector in SW Africa, Kurt Dinter. He records that he found specimens on the northern spurs of the Auas Mt on dry quartz koppies and also near Grootfontein in deep black turf which he considered remarkable adaptation to different ecological conditions. He pointed out on other occasions that most species of the genus he had collected showed considerable vegetative variability and that great care was necessary in selecting characters for the distinguishing of species.

It was the recent receipt of fresh flowers in spirit of *Brachystelma dinteri* from Peter Bruyns which revealed that I had erred in the Flora of S.Afr. 27,4 (1980) in placing under it *B.brevipedicellatum* and *B.ringens.* Although the appearance of the flowers is very similar there is a distinction in the coronal structure.

This revised concept of *B.dinteri* restricts its distribution to the northern parts of SW Africa and *B.brevipedicellatum* is treated under the following heading no.43A.

43A. **Brachystelma brevipedicellatum** *Turrill* in Kew Bull. 1922:29 (1922); under *B.dinteri* in error, R.A.Dyer in Flora S.Afr. 27,4:26 (1980)
Brachystelma ringens E.A.Bruce in Flower.Pl.Afr. 28:5.1096 (1951), under *B.dinteri* in error R.A.Dyer l.c. (1980) [Fig.43A, 43B]

Perennial herb with tuber. *Tuber* 50-60 mm diameter, about 30 mm thick, compressed from above. *Stems* 2-4, produced annually, up to about 100 mm tall, erect or often branched and somewhat spreading, thinly covered with reflexed hairs. *Leaves* elliptic to ovate-oblong or ovate-lanceolate, 15-50 mm long, 10-20 mm broad, slightly folded upwards, sometimes crisped on margin, hairy on lower surface and near margin on upper surface. *Flowers* few to several together, shortly stalked, in the axils or somewhat lateral at the nodes. *Corolla* opening more or less flat, about 10 mm diameter with a small central tube or depression less than 2 mm deep containing the corona divided about halfway, pale green outside, pale maroon near base within; lobes ovate-deltoid,

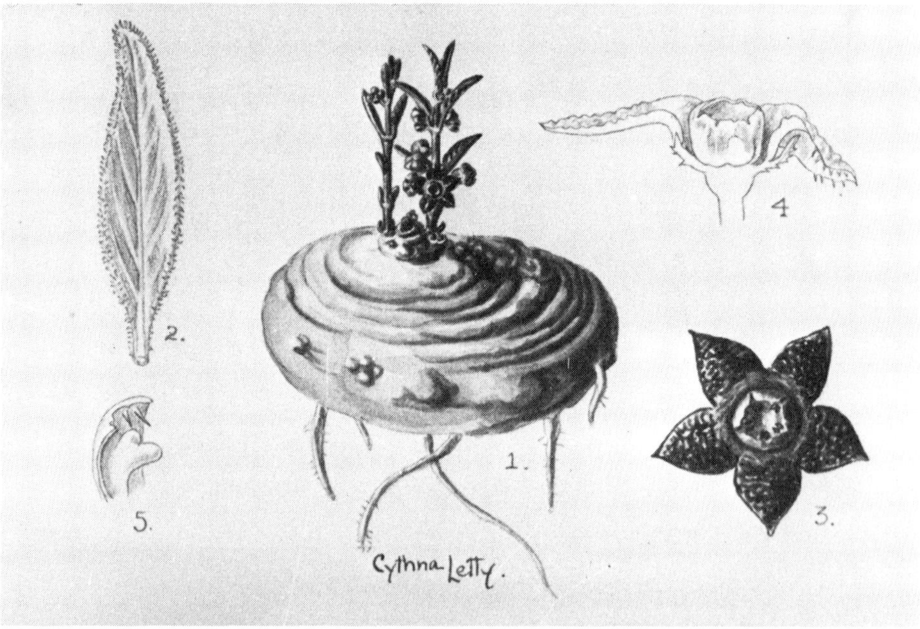

Figure 43A. *Brachystelma brevipedicellatum:* copy of plate by Cythna Letty in Flower.Pl.Afr. 28:Plate 1096b (1951) (as *B.ringens*)

Figure 43B. *Brachystelma brevipedicellatum:* a. (*B.ringens* E.A.Bruce) tuber with first stem growth of season producing flowers clustered at nodes, x 1; b. single stem in fruit from old stem and in flower on new growth, x 1 (*D.S.Hardy* 2205; photo: J.Reyburn); c. flower face view showing cupular outer corona with slits in margin opposite the nectariferous pockets, x 4 (*E.A.Phillips* 119, photo: H.King)

transversely rugose or corrugated on surface, dark maroon, glossy, sometimes with few hairs.

Brachystelma brevipedicellatum was given its name because of the very shortly stalked flowers. It was described from a plant cultivated at the Royal Botanic Gardens, Kew, which had been received from Pretoria about 1920 but apparently without any field record. To complicate matters further, the author, Turrill, states that there is no outer corona which there is and he describes it as the inner corona. The interpretation arrived at indicates that there is no specific difference between it and *B.ringens* E.A.Bruce, which name is to be regarded as a synonym.

The distribution of *B.brevipedicellatum* seems to range from the Vryburg district of the north-eastern Cape extending into warm areas of the north-eastern Transvaal in the vicinity of Phalaborwa and Tzaneen, avoiding the cold highveld.

In the Flora of Southern Africa 27,4:26 (1980) I misinterpreted the structure of the corona of *B.dinteri* (43) with the result that I regarded *B.brevipedicellatum* as falling within its specific range.

The following comparison in key form summarises the differences between the closely related species *B.dinteri* and *B.brevipedicellatum,* which may be followed further in the figures.

Stems erect sparsely branched from near base; leaves hairy on both surfaces; corona saucer or broadly cup-shaped at base giving rise to 5 broad emarginate or toothed lobes incurved and overtopping staminal column, confluent at base with small inner lobes; inner lobes incumbent on base of anthers **B.dinteri**

Stems erect or spreading from base; leaves hairy on under surface and on margin of upper surface; corona cup-shaped with incurved margin overtopping staminal column, with 5 slits opposite the pollinia and sometimes 5-notched behind the inner lobes; inner lobes small, arising from within base of outer tube and incumbent on base of anthers **B.brevipedicellatum**

B.brevipedicellatum is an attractive little plant in cultivation because of its clusters of dark maroon flowers at the nodes of the stems, usually two or more being open at the same time.

44. **Brachystelma schinzii** *(K.Schum.) N.E.Br.* in F.T.A. 4,1:471 (1903); R.A. Dyer in Flora S.Afr. 27,4:27 (1980) [Fig.44]

Perennial herb with tuber. *Tuber* about 30 mm diameter, conical-ovoid. *Stems* 1-few produced annually, not or only sparingly rebranched, 80-100(150) mm tall. *Leaves* narrow up to 50 mm long, sometimes with margins recurved and appearing subterete. *Flowers* 1-2 lateral at the nodes, on thin stalks 12-16 mm long. *Corolla* spreading up to 10 mm diameter, lobed to about two-thirds way, purple-brown, with white base round corona; tube absent or short and sur-

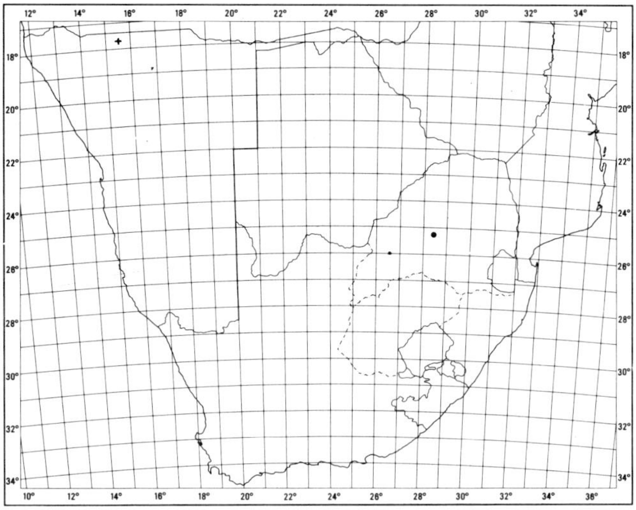

\+ *Brachystelma schinzii* (44)
● *Brachystelma discoideum* (45)

Figure 44. *Brachystelma schinzii:* portion of flowering plant at BRI, showing characteristic long hairs pendulous from the corolla and the disc-like corona, x 2 (*B.de Winter* 6877, from Ovamboland)

rounding corona; lobes about 4 mm long and 2 mm broad at base, narrowed to apex, long hairy within with subclavate (vibratile?) purple hairs.

Brachystelma schinzii is named after the Swiss botanist Hans Schinz who made an extensive botanical journey in southern Africa early in the present century. Unlike *Brachystelma dinteri* and *B.cupulatum* which also occur in northern SW Africa, and have a wide distribution, *B.schinzii* seems very much more restricted in the northern parts of Ovamboland. The long subclavate purple hairs on the inner surface of the corolla protrude from the shrivelled old corollas unlike any other of the small-flowered species. Another feature of special interest is the thickened disc-like outer corona and cushion-like inner lobes which relate it fairly closely with *B.discoideum* known only from the Transvaal north of Pretoria. Our record of the species is based on *De Winter & Giess* 6877. Two tubers came into flower at the same time at BRI, and showed considerable variation in character. Leaves varied in width and the degree of revolution of the margins. One (a) had long hairs round the corona on the inner face of the corolla, the other (b) was without hairs on the face; the former had a sunken corona in the central corolla depression, the other had the corona ± exposed and notched on the margin between the inner lobes (b) which is illustrated in the figure. When the flowers close with age the hairs protrude like long whiskers.

45. **Brachystelma discoideum** *R.A.Dyer* in Flower.Pl.Afr. 42:t.1668 (1973); Flora S.Afr. 27,4:27 (1980) [Fig.45]

Perennial herb with tuber. *Tuber* up to about 70 mm diameter, considerably depressed. *Stems* 1-2(4) produced annually, unbranched or sparingly branched, up to 100 mm tall. *Leaves* shortly stalked; blade 5-15(25) mm long, 5-15 (20) mm broad. *Flowers* 1-2 lateral at nodes, on slender stalks 8-12(15) mm long, slightly decurved. *Corolla* spreading, star-shaped, light yellow, 1,8-2,2 cm diameter, divided nearly to base with central 1,5 mm slightly raised in the middle into a small cushion; lobes slender, 8-10 mm long with margins recurved.

Brachystelma discoideum has an unusual corona for the genus, disc-like and resembling that of *B.schinzii.* Unfortunately it is known only from one collection when Dave Hardy discovered two tubers near Salt Pan about 40 km north of Pretoria in gravelly sandy soil. Only one of these tubers survived at BRI and served to illustrate the species in 1971. If anyone might be expected to discover additional material it would be Dave Hardy, but in spite of several excursions arranged for that specific purpose there were no positive results. The accompanying figure shows the distinctive corona clearly and at the same time does justice to the star-like corolla which is divided nearer to the base than in most other species.

46. **Brachystelma pygmaeum** *(Schltr.) N.E.Br.* in Fl.Cap. 4,1:857 (1908); R.A.Dyer in Flora S.Afr. 27,4:28 (1980) [Fig.46, 46A]

Perennial dwarf herb with tuber. *Tuber* subglobose, somewhat depressed, 40-100 mm diameter. *Stems* few to many produced annually, sometimes the

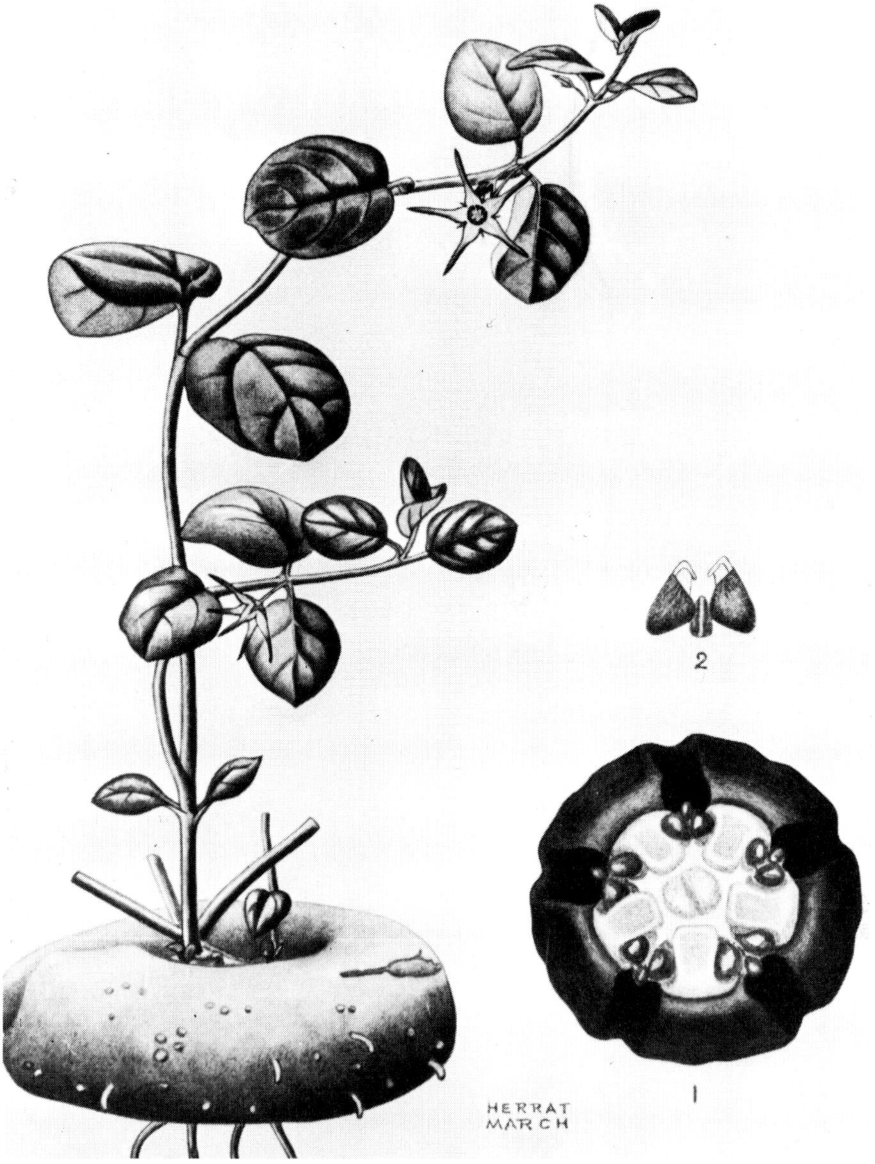

Figure 45. *Brachystelma discoideum:* the type plant as illustrated in Flower. Pl. Afr. t.1668, x 1; 1. the corona from above showing 5 pockets and the rest disc-like round the staminal column, x 20; 2. pollinia, x 50 (*D.Hardy* 2440 north of Pretoria)

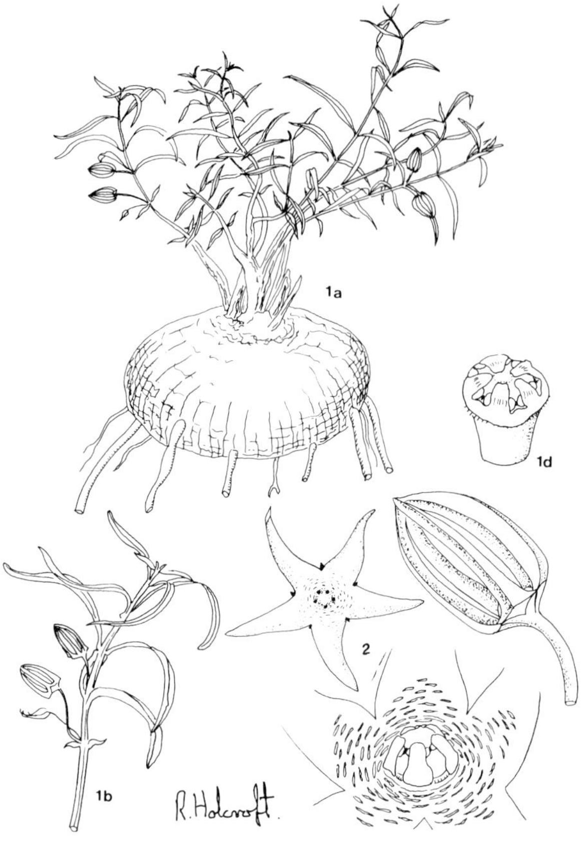

Figure 46. *Brachystelma pygmaeum* subsp. *pygmaeum:* 1a. plant, slightly reduced; 1b. flowering branch, x ½; 1c. flower with corolla lobes united at the tips forming cage, x 2; 1d. staminal column with corona much reduced, arising from near top and represented by a ring of teeth, x 5; 2. subsp. *flavidum:* flower open with corolla lobes free and with medifixed hairs round corona, x 2; 2b. united portion of corolla showing medifixed hairs round corona, x 5 (after Flower. Pl. Afr. t.1067, 1088 Flora S.Afr. 27,4: t.5 (1979))

flowers appearing before the leaves, simple or sparingly branched, erect or spreading with age, 40-80 mm long. *Leaves* 5-20 mm long, up to 6 mm broad, narrowed into short stalk, sometimes folding upwards with age. *Flowers* 1-3 together, opening in succession on stalks 3-8 mm long, with or without medifixed hairs round the corona.

Brachystelma pygmaeum so named because of its dwarf habit is a constituent of the spring-flowering perennial herbs in grassveld which respond to early veld fires, even to the condition where the flowers precede the leaves on occasions. The columnar staminal column with the corona represented by a toothed rim is unique in the genus as also the medifixed hairs which more often than not occur in circles round the corolla base. Each hair is attached to the corolla by the middle and the two arms spread closely on the surface where they may readily be overlooked.

The species is subdivided into two subspecies based on the following characters:

Figure 46A. *Brachystelma pygmaeum* subsp. *flavidum:* plant collected near Dumisa in southern Natal grassveld, x 2 (*A. W.Bayer* PRE 27156, cult. BRI, flowered Oct., photo: R.A.Dyer)

Corolla-lobes united at the tips into a cage-like structure, rarely some lobes breaking free normally – subsp. **pygmaeum**
Corolla-lobes free at tips and expanding star-shaped – subsp. **flavidum**

subsp. **pygmaeum** *R.A.Dyer* in Flora S.Afr. 27,4:28 (1980) [Fig.46]
Corolla with short tube and lobes united at tips, rarely with some free; tube up to 1,5 mm long with or sometimes without medifixed hairs round corona and on base of lobes.

Brachystelma pygmaeum subsp. *pygmaeum* extends from the eastern Cape into the grassveld of Natal and Zululand and is not infrequent in the Transvaal grassveld. In this subspecies the corolla-lobes are normally united but are quite often to be seen free in herbaria. In these cases they have probably had some rather rough handling during preparation which has broken the delicate union of the tips. During field work, Fanie Venter of BRI staff, made a special observation on 2 August 1978, that near Angle Station on the road from Barberton to Havelock he found large populations of *B.pygmaeum* subsp. *pygmaeum.* He noted that most of the plants had tubers completely covered by quartzite soil, whereas in others the tubers were wedged between rocks, and somewhat exposed: further that the only populations in flower were in burnt grassveld and in these all corolla-lobes were united at the tips in the field.

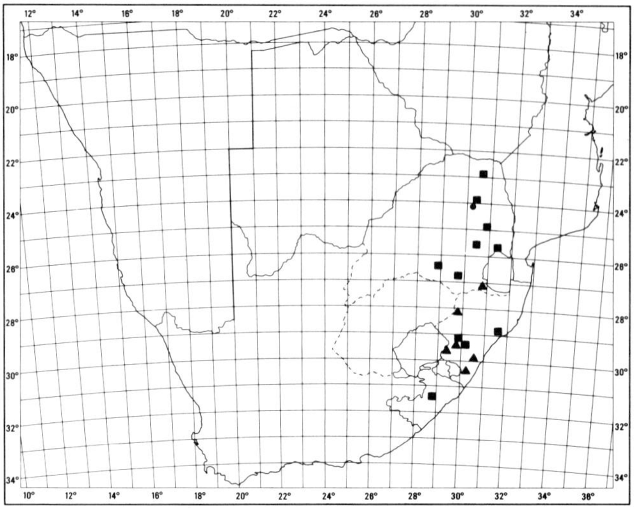

■ *Brachystelma pygmaeum* subsp. *pygmaeum*
▲ *Brachystelma pygmaeum* subsp. *flavidum*
● *Brachystelma pachypodium*
(47)

subsp. **flavidum** *(Schltr.) R.A.Dyer* in Flora S.Afr. 27,4:28 (1980) [Fig.46A]
Corolla lobes free at the tips, spreading from a more or less flat base, with or without medifixed hairs surrounding the corona.

Brachystelma pygmaeum subsp. *flavidum* was at first given specific rank as *B. flavidum* and named because of the yellowish colour of the corolla. But the distinguishing character of the corolla-lobes being free and not united at the delicate tips was found to be too delicate a difference. The subspecies originated in southern Natal and extends over a similar distribution range as that for subspecies *pygmaeum* to the Transvaal, but has not been recorded in the Transkei.

47. **Brachystelma pachypodium** *R.A.Dyer* in Flower.Pl.Afr. 32:t.1269 (1958); in Flora S.Afr. 27,4:28 (1980) [Fig.47]

Perennial herb with tuber. *Tuber* relatively large, up to 130 mm diameter, slightly depressed, about 100 mm thick, irregularly shaped and sometimes partly exposed above ground. *Stems* 1-several, produced annually, suberect or spreading, 70-200 mm long. *Leaves* shortly stalked, 25-50 mm long, 20-40 mm broad. *Flowers* 1-2 lateral at the nodes sometimes on both sides of petiole. *Corolla* bright greenish-yellow, 20-23 mm long, divided three-quarter of length, with tips of lobes insecurely united, or sometimes free; tube bell-shaped, about 5 mm long, 8 mm diameter; lobes slender, about 17 mm long with the margins pinched inwards above base and slightly undulate above.

Brachystelma pachypodium seems to have a restricted distribution in shady

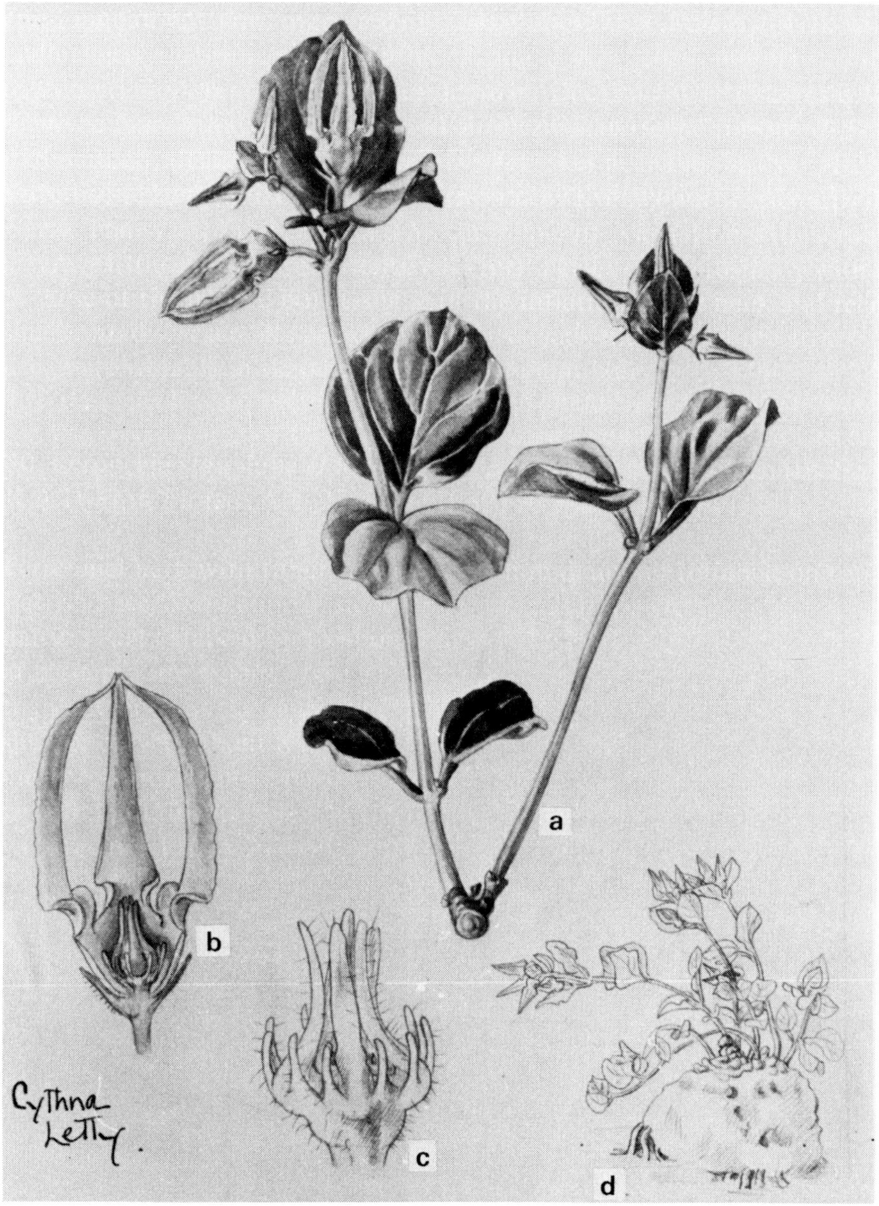

Figure 47. *Brachystelma pachypodium:* a. flowering branch, x 1; b. longitudinal section of flower, x 2,5; c. corona, with very deeply divided outer lobes and linear-erect inner lobes, x 9; d. habit, showing massive tuber (photo of plate 1269 in Flower. Pl. Afr. 1958)

kloofs in the mountains of The Downs in the northern Transvaal. The bright greenish-yellow flowers and the comparatively large fleshy or 'stout' tubers often exposed above ground are so distinctive that they could not long be overlooked in any explored locality. The pinched-in base of the corolla-lobes is a unique feature among the related species. The fusing of the tips of the corolla-lobes is so delicate that they may become free even with gentle handling.

48. **Brachystelma gracile** *E.A.Bruce* in Flower.Pl.Afr. 27:t.1077 (1949); R.A. Dyer in Fl.S.Afr. 27,4:30 (1980) [Fig.48]

Perennial herb with tuber. *Tuber* 60-100(150) mm diameter, depressed. *Stem* produced annually, simple or occasionally branched from base, rarely above, slender, up to 700 mm high, internodes 20-60 mm long. *Leaves* with practically no stalk, very slender, 40-90(120) mm long, 1-2(5) mm broad, margins strongly incurved. *Flowers* 1-4, axillary or lateral at nodes, developing successively on slender stalks 7-10 mm long. *Corolla* 10-17 mm long, with slender

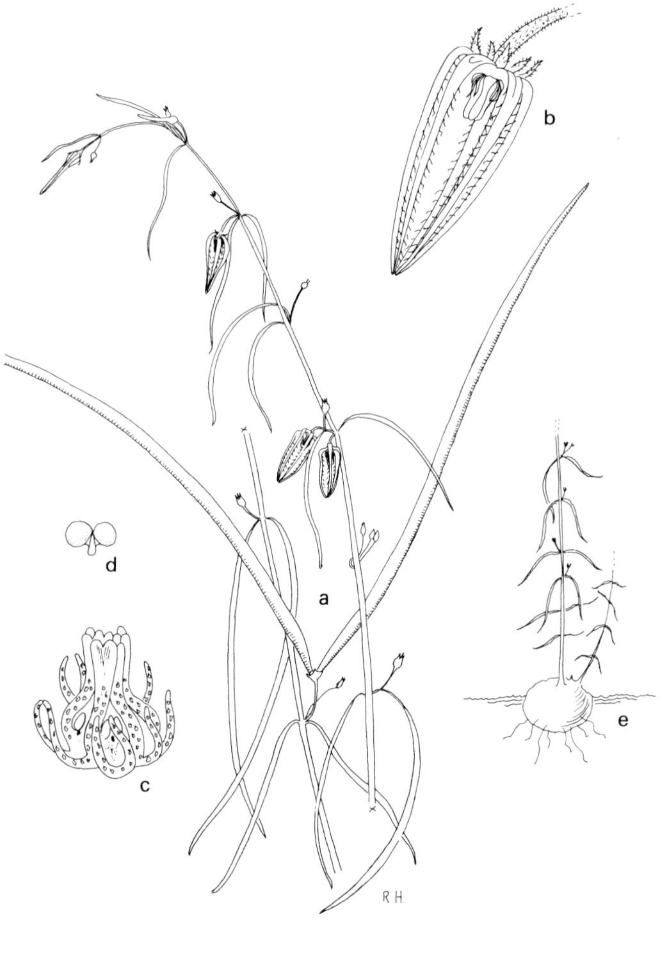

Figure 48. *Brachystelma gracile:* a. stem with fruit (follicles) and flowers, x ¾; b. flower, x 3; c. x 8; d. pollinia, x 15; e. habit (after Edith Burges in Flower. Pl. Afr. 5. 1077 (1949))

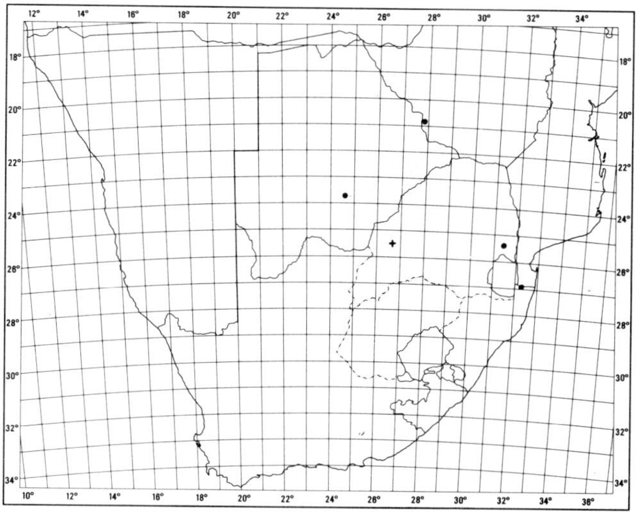

● *Brachystelma gracile* (48)
+ *Brachystelma gracillimum* (49)

beak in bud, lobed almost to base; lobes united at tips, very slender with recurved margins and with long soft white hairs.

Brachystelma gracile is notable for its 'slender' habit and an unusual distribution record. The original discovery was of a plant near Plumtree in Rhodesia in 1944. In subsequent years it has been collected in Botswana, three times in Transvaal and twice in Zululand. This wide sporadic distribution suggests that the species is particularly adaptable to new environments and has spread southwards within recent times. Like most others of the genus the tubers are edible and sought after by indigenous peoples.

49. **Brachystelma gracillimum** *R.A.Dyer* in Jl S.Afr.Bot. 11:112 (1945); in Flora S.Afr. 27,4:30 (1980) [Fig.49]

Perennial herb with tuber. *Tuber* 30-50 mm diameter. *Stems* 1-2 produced annually, up to about 450 mm tall, with basal internodes up to 100 mm long, moderately branched above from alternating nodes. *Leaves* shortly stalked, up to about 15 mm long and 7,5 mm broad. *Flowers* solitary, lateral at nodes. *Corolla* about 45 mm long, with long slender beak, lobed almost to base; lobes united at tips, very slender, about 0,5 mm broad not opening into a cage.

Brachystelma gracillimum has a conspicuously slender corolla to justify its name. It was described before *B.gracilis* (sp.48) which has since been shown to have a wide distribution in southern Africa. *B.gracillimum* on the other hand has not been recorded more than the once in the Marico district of the western Transvaal, having been discovered there 40 years ago by W.J.Louw. Its floral structure including the unusual corona is noteworthy.

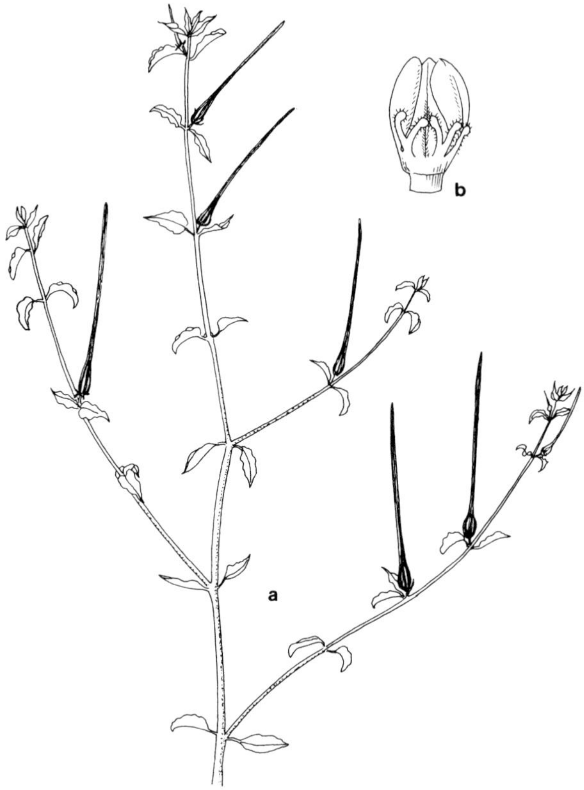

Figure 49. *Brachystelma gracillimum:* a. top portion of plant in flower, x ⅔; b. corona with outer lobes deeply bilobuled, inner lobes oblong-elliptic, most unusual in size and shape, x ⅔; (after *J.Louw* 811 in Jl S.Afr. Bot. 11:112 (1945))

50. **Brachystelma gemmeum** *R.A.Dyer* in Bothalia 12,4:628 (1979); in Flora S.Afr. 27,4:30 (1980) [Fig.50]

Perennial dwarf herb with tuber. *Tuber* about 60 mm diameter. *Stems* 4-5 produced annually, 5-7 mm tall. *Leaves* dense, shortly stalked, 30-40 mm long, 3-4 mm broad, tapering to both ends. *Flowers* few-several, congested lateral to the nodes, developed successively. *Corolla* 15-16 mm long, purple, divided nearly to base; tube bell-shaped, about 2 mm long, slightly spreading at base of lobes, with few 3,5 mm long narrowly cylindric, vibratile, caducous, purple hairs on inner face at base of lobes; lobes slender, 13-14 mm long, margins recurved, tips united and forming slender cage.

Brachystelma gemmeum is the most recent new species discovery within the genus. It was found in December 1977 by Percy Lancaster, near the Gravelotte Emerald Mine in the Phalaborwa district of the north-eastern Transvaal. It is claimed to be particularly rare on the evidence that a team of mine workers set to hunt for tubers found very few. It is unique among its relatives in the presence of 3,5 mm long, narrowly cylindric, vibratile, purple hairs in the throat of the corolla. These hairs respond to a slight movement of the surrounding air by vibrating from the base and readily become detached with age. It is confidently interpreted that the movement of the hairs assists in attracting pollinating insects. The name *gemmeum* was chosen because of the association of its discovery with the Gravelotte Emerald mine. Unfortunately no fresh living material was available for illustration and the reader is left to do a little imaginative reconstruction for himself on the limited data presented.

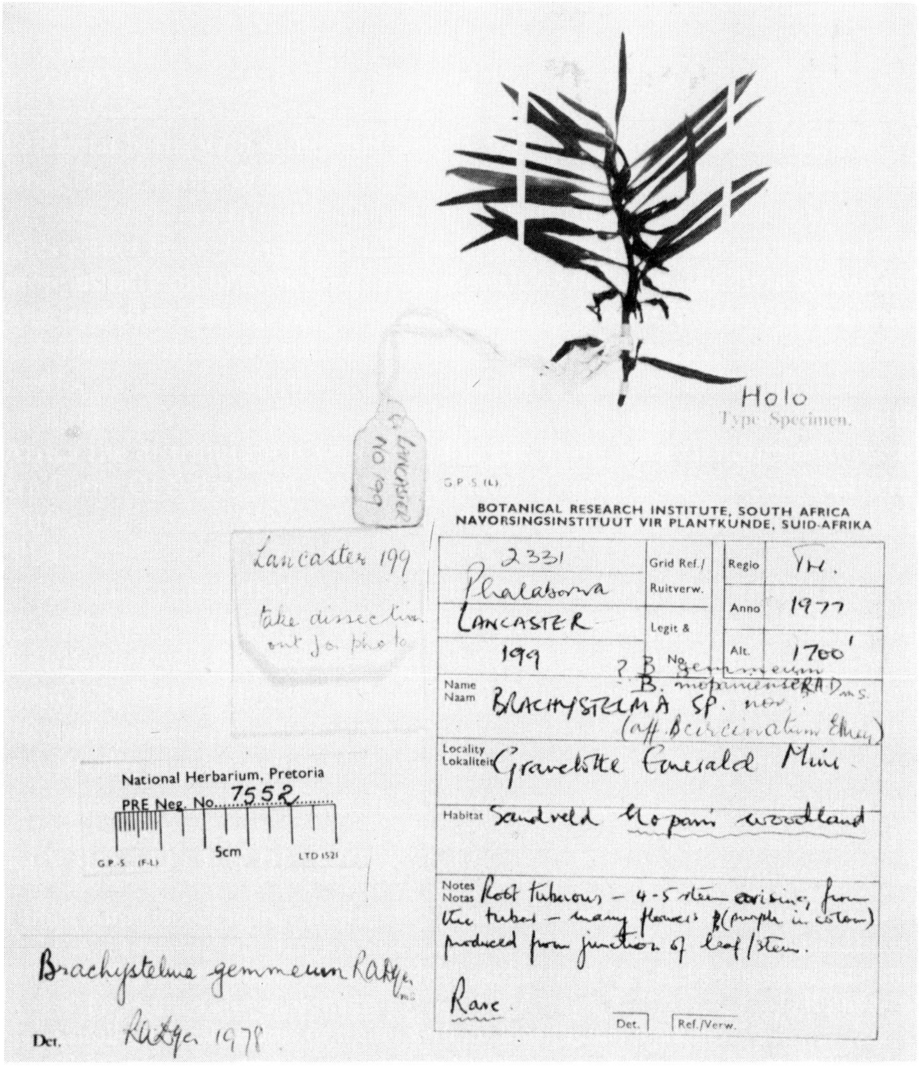

Figure 50. *Brachystelma gemmeum:* photo. of type specimen, unfortunately the diagnostic 3,5 mm long, narrowly cylindric, vibratile hairs in the corolla throat are not seen distinctly, x 1 (*Percy-Lancaster* 199)

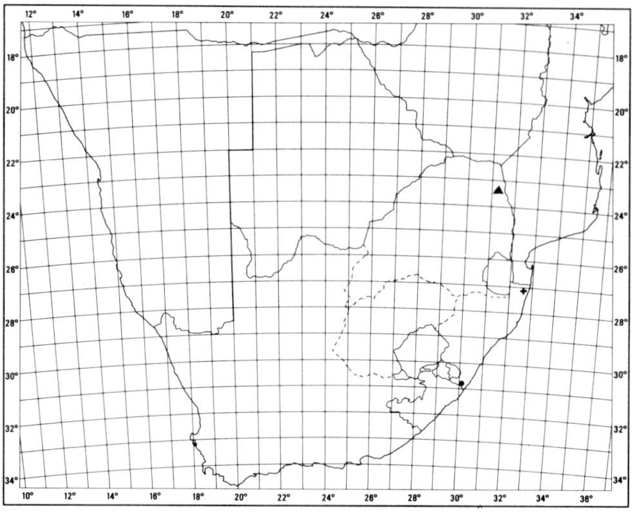

▲ *Brachystelma gemmeum* (50)
● *Brachystelma tenellum* (51)
\+ *Brachystelma tenue* (52)

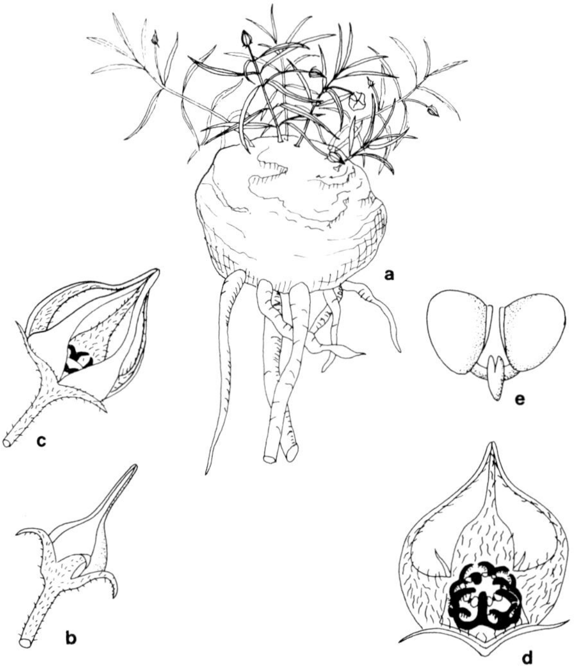

Figure 51. *Brachystelma tenellum:* a. plant, x ½, sometimes more densely branched; b. bud with characteristic long slender beak and long recurving sepals (calyx segments), x 4; c. open flower with tips of corolla lobes united, glabrous outside, hairy within, x 4; d. corolla with two lobes removed exposing corona, bilobuled outer lobes and incumbent inner lobes on staminal column, x 5; e. pollinia attached to carrier, x 20 (after *Strey* 10050 Flower. Pl. Afr. t.1664)

51. **Brachystelma tenellum** *R.A.Dyer* in Flower.Pl.Afr. 42:t.1664 (1973); in Flora S.Afr. 27,4:31 (1980) [Fig.51]

Perennial dwarf herb with tuber. *Tuber* 40-50 mm diameter depressed. *Stems* 1-few produced annually, slender, unbranched or sparsely branched near base, suberect or spreading, 20-50 mm long. *Leaves* very shortly stalked, slender, 10-20 mm long. *Flowers* single lateral at the nodes on slender stalks 10-15 mm long. *Corolla* 3,5-4 mm long with short slender beak, divided almost to base; lobes united at tips, attenuate from slightly above base, purple.

Brachystelma tenellum has a very restricted known distribution on Oribi Flats in southern Natal which is inland from Port Shepstone on Gibralter and Highlands farms. It has been collected here on several occasions in pockets of weathered Table Mountain sandstone but nowhere further afield. The sharp buds and miniature cagelike flowers are particularly delicate and attractive in their small way, which is better appreciated under magnification. R.Strey, retired from BRI, made the original discovery of the species in November 1970, and has been associated with all subsequent collections up to now.

52. **Brachystelma tenue** *R.A.Dyer* in Bothalia 10:376 (1971); in Flora S.Afr. 27,4:31 (1980) [Fig.52]

Perennial herb with tuber. *Tuber* 40-50 mm diameter, somewhat depressed, red. *Stems* 1-few produced annually, spreading or erect, up to about 100 mm tall densely covered with spreading transparent hairs up to 1 mm long. *Leaves* shortly stalked, 10-50 mm long, 1,5-15 mm broad. *Flowers* 1-2, lateral at nodes on slender stalks 8-13 mm long and thickening only after fertilization. *Corolla* yellowish-brown, 15-20 mm long, divided nearly to base, with lobes united at tips, tube saucer-shaped 1-2 mm deep; lobes slender 14-18 mm long.

Brachystelma tenue by comparison with related species seems rather featureless and the main interest is to study the relationship between it and *B.circinatum.* The typical form was derived from the Ubombo district near the coast at Lala Nek. A second specimen from near the road from lower Mkuze to Bazwane was more robust and a further specimen from near Nongoma leaves one in two minds as to its correct classification. It is most unsatisfactory to have to come to conclusions on inadequate material and no field work and so an answer to this last problem is being left in abeyance. The original specimen described was collected by Hans Vahrmeijer in 1965 and it was only later that the similar specimens collected earlier near Nongoma by Gerstner were located in the Natal Herbarium, Durban.

53. **Brachystelma villosum** *(Schltr.) N.E.Br.* in F.C. 4,1:863 (1908); R.A.Dyer in Flora S.Afr. 27,4:31 (1980) [Fig.53]

Perennial herb presumably with tuber. *Tuber* (unseen) with stout underground neck. *Stems* 1-several, produced annually from near base, 100-300 mm tall

Figure 52. *Brachystelma tenue:* a. typical form preserved in the National Herbarium, Pretoria, x ½; b. part of 'a' enlarged showing developing fruits and thread-like corolla lobes, x ½ (*J. Vahrmeijer* near Lala Nek on Zululand coast on Ubombo-Ingwavuma border)

with long dense, soft, spreading hairs. *Leaves* 10-25 mm long, 5-7 mm broad. *Flowers* 2-3, lateral at nodes on short stalks. *Corolla* cage-like, 20-30 mm long; tube 4-5 mm long; narrowly bell-shaped; lobes 12-25 mm long united at apex, forming a cage-like body.

Brachystelma villosum appears to have a distribution range in grassveld from the northern Transvaal in the Pietersburg district extending to the eastern Transvaal, Swaziland and northern Natal. The original collection by E.E.Galpin was made in 1889 on the grassy hills above Barberton in the eastern Transvaal. Galpin makes the statement in his records that the flowers are sweetly scented, which is unexpected but not unique in a genus noted for some very

Figure 53. *Brachystelma villosum:* the typical form, preserved in the National Herbarium, Pretoria, rarely recorded, x ⅔ (*Galpin* 588, near Barberton, eastern Transvaal)

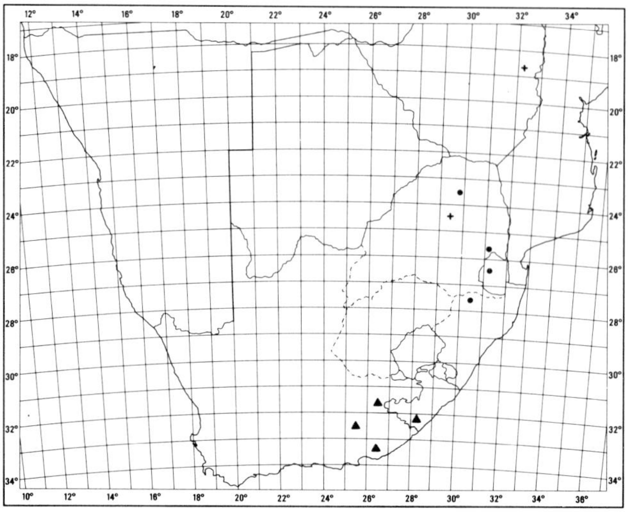

● *Brachystelma villosum* (53)
\+ *Brachystelma pilosum* (54)
▲ *Brachystelma elongatum* (55)

bad smelling flowers such as *B.foetidum.* The species was in fact first described by Schlechter under the generic name *Dichaelia* because of the united tips of the corolla-lobes but as mentioned earlier this distinction was later abandoned but the specific name *villosum* meaning covered with long weak (soft) hairs was retained.

54. **Brachystelma pilosum** *R.A.Dyer* in Bothalia 6:541 (1956); in Flora S.Afr. 27,4:32 (1980) [Fig.54]

Perennial herb with tuber (presumed). *Stem* produced annually, up to about 250 mm tall, sparingly branched, with copious spreading hairs and internodes 20-40 mm long. *Leaves* 10-15 mm long, 3-5 mm broad, with margins somewhat folded upwards and slightly undulate. *Flowers* usually 2 lateral at the nodes. *Corolla* 16-18 mm long, shortly united at base; tube saucer-shaped, 1-1,5 mm deep; lobes united at tips, narrowed to about 0,5 mm in the middle.

Brachystelma pilosum a softly hairy one, by comparison with most other species, might deserve the term drab. It is unlikely to be seriously sought after by collectors, even though the internal structure of the flower is just as complex and beautiful as its relatives. The type collection was made in 1931 from near Naboomspruit in the northern Transvaal by E.E.Galpin and an earlier one was from near Potgietersrus in wooded grassveld by V.G.Crawley. A specimen sent by D.Plowes from eastern Rhodesia, seems specifically equal which extends the distribution range appreciably north.

Figure 54. *Brachystelma pilosum:* the typical form, preserved in the National Herbarium, Pretoria, x ½ (*E.E. Galpin* 11575, near Naboomspruit, north of Pretoria)

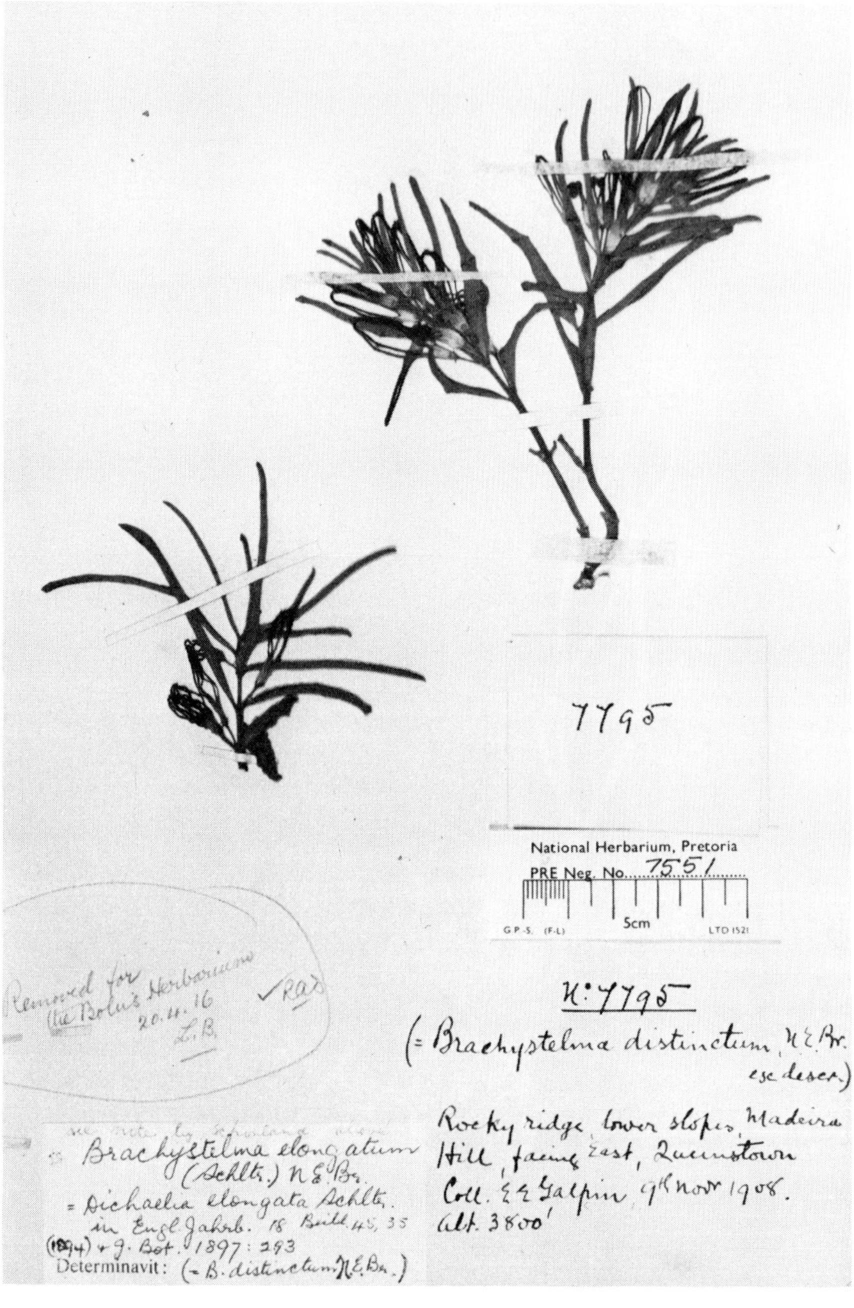

Figure 55. *Brachystelma elongatum:* typical form preserved in the National Herbarium, Pretoria, x $^2/_3$ (*E.E. Galpin* 7795 from Queenstown, eastern Cape)

55. **Brachystelma elongatum** *(Schltr.) N.E.Br.* in F.C. 4,1:862 (1908); R.A. Dyer in Flora S.Afr. 27,4:32 (1980) [Fig.55]

Perennial herb with tuber. *Tuber* 50-90 mm depressed. *Stems* 1-3 produced annually, up to 150 mm tall, shortly and softly hairy, moderately branched towards base, leafy. *Leaves* 7-30 mm long, up to 8 mm broad, hairy. *Flowers* 2-5 together lateral at nodes; mostly open at same time. *Corolla* 15-25 mm long, shortly tubular; tube up to about 5 mm long, 4 mm wide at mouth; lobes united at tips, 12-20 mm long, very slender with recurved margins, hairy outside, smooth within.

Brachystelma elongatum seems at one time to have been relatively frequent in the eastern Cape Province but there are few modern collections. One variant was given the name *Brachystelma distinctum,* but in the light of additional material it could not be upheld. The long cage-like corollas are attractive in their abundance. If collectors wish to search for more records this should be done anywhere in the grassveld of the Somerset East district of the eastern Cape, extending eastwards into the Transkei. This is another species first described by Schlechter in 1894 in the now abandoned genus *Dichaelia* because of the united tips of the corolla lobes.

56. **Brachystelma dimorphum** *R.A.Dyer* in Bothalia 12,4:627 (1979); in Flora S.Afr. 27,4:32 (1980) [Fig.56, 56A]

Perennial herb with tuber. *Tuber* 50-70 mm diameter, bell-shaped and rounded at base or somewhat depressed. *Stem* single (2) developed annually, sparingly

■ *Brachystelma dimorphum* subsp. *dimorphum*
◆ *Brachystelma dimorphum* subsp. *gratum*
● *Brachystelma circinatum* (57)
▲ *Brachystelma stenophyllum* (58)

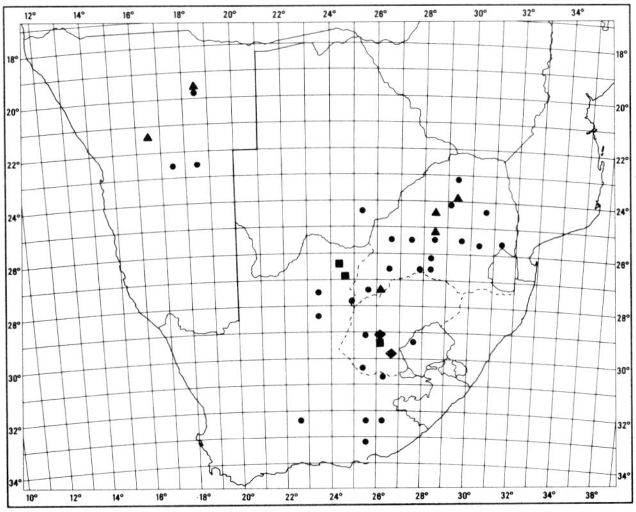

Figure 56. *Brachystelma dimorphum* subsp. *dimorphum:* a. habit, x ½; b. flowering twig with bud, x 2; c. flowering twig with flowers and bud, corona exposed, without outer lobes produced, x 2 (after *R.Hutchinson*, from near Vryburg, photo Adela Romanowski)

branched, 10-20 mm high. *Leaves* shortly stalked, 5-20 mm long, 4-14 mm broad, folded upwards, somewhat hairy on outer surface and smooth within. *Flowers* 1-few, lateral at nodes, opening in succession. *Corolla* 6-7 mm long with cylindric beak in bud, divided almost to base, without hairs on outer surface with few hairs on inner surface.

Brachystelma dimorphum was recognised as a distinct species during a close examination of the material of *B.circinatum* at BRI. It was found that a few specimens did not fall within the circumscription of that admittedly very variable species. The distinctions of *B.dimorphum* are that the outer corona-lobes are almost obsolete or the outer margin very shortly bidentate and the inner surface of the corolla-lobes are shortly hairy in the lower half, while the outer surface is usually without hairs. It was found at the same time that the corolla occurred in two forms, hence the name and that of the two subspecies should be recognised on the following features:

Corolla-lobes united at tips into a cage-like structure; inner corona-lobes short, subquadrate subsp. **dimorphum**

Corolla-lobes free at tips, spreading, star-shaped, recurved; inner corona-lobes slightly longer than broad subsp. **gratum**

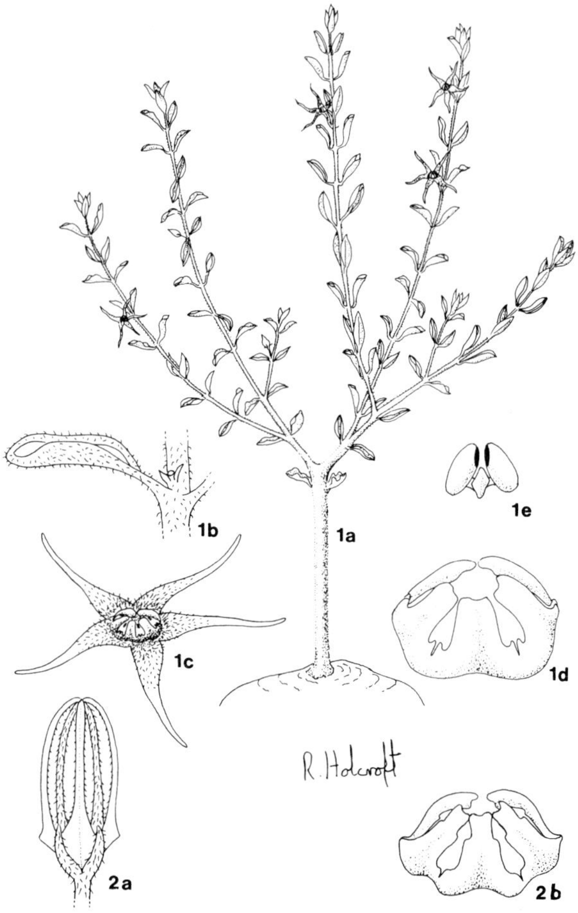

Figure 56A. 1. *Brachystelma dimorphum* subsp. *gratum:* 1a. habit, x ½; 1b. leaf with characteristic half-folded habit, and hairy on outer surface, x 3; 1c. flower with widely spreading lobes hairy on inner surface, outer margin of pockets not produced into lobes, x 3; 1d. corona without produced outer lobes, inner lobes incumbent on staminal column, truncate, x 10; 1e. pollinia shortly attached to carrier, x 20 (after *Rawlinson* in PRE 57440); 2. *Brachystelma dimorphum* subsp. *dimorphum:* 2a. flower with corolla-lobes united at tips, glabrous outside, hairy inside, x 3; 2b. corona showing slight variation from 1d, x 10 (after voucher in BRI)

subsp. **dimorphum** *R.A.Dyer* in Flora S.Afr. 27,4:33 (1980) [Fig.56]
Brachystelma dimorphum subsp. *dimorphum* has been traced in its distribution by the specimens of several collectors to occur from the Vryburg district of the north-eastern Cape to Kimberley and over the border into the Orange Free State near Bloemfontein. The tubers show a preference for alluvial soil in expansive shallow pans.

subsp. **gratum** *R.A.Dyer* in Flora S.Afr. 27,4:33 (1980) [Fig.56A]
Brachystelma dimorphum subsp. *gratum* is a similar case to *Brachystelma pygmaeum* and its subspecies *flavum* (sp.46), the distinction between free and united corolla-lobes. At times in the past this distinction was regarded as of generic importance but this viewpoint had to be abandoned because of its instability. *Brachystelma pachypodium* is an example of a very delicate connection between the tips of corolla-lobes which is very readily severed.

The first record of subsp. *gratum* was made by S.Rawlinson near the Welkom aerodrome in an alluvial pan very similar to the conditions for subsp. *dimorphum*. The tuber flowered in Bloemfontein and continued to do so in BRI.

57. **Brachystelma circinatum** *E.Mey.* Comm. 196 (1836); R.A.Dyer in Flora S.Afr. 27,4:33 (1980) [Fig.57, 57A, B, C, D]

Perennial herb with tuber. *Tuber* up to 100 mm diameter, depressed and often concave above. *Stems* 1-few produced annually, 70-300 mm high, usually branched above. *Leaves* 5-15(20) mm long, very variable in width, usually hairy

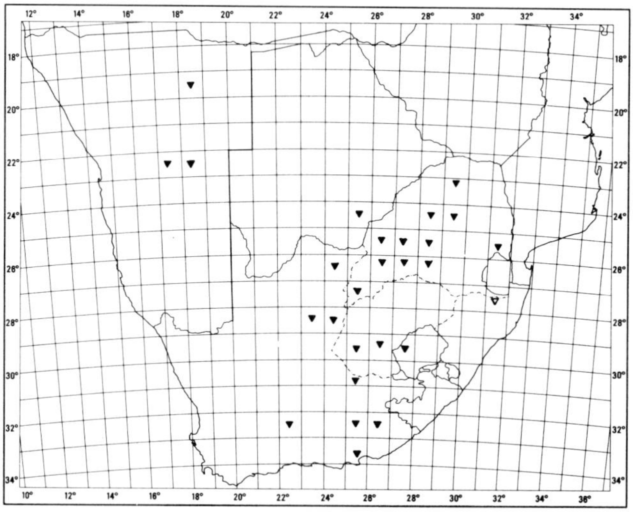

▼ *Brachystelma circinatum* (57)

Figure 57. *Brachystelma circinatum:* a. plant first named *B. ovatum,* with few annual stems from tuber, x 1; b. flower with corolla lobes united at tips, x 2; c. two-seriate corona with the outer lobes deeply bilobuled, inner lobes incumbent on anthers, x 5; d. pollinia and carrier (after Ref. Bot. 4:t.226)

on both surfaces, sometimes without hairs on upper, flat or undulate sometimes upfolded. *Flowers* 1-few at nodes. *Corolla* from white to maroon, divided nearly to base, minutely hairy outside without hairs inside; tube 0 or slightly saucer-shaped; lobes united at tips, slender, very variable in length.

Brachystelma circinatum is a composite species of diversified vegetative forms. The distribution stretches from northern SW Africa, Botswanaland and Transvaal southwards to the eastern Cape, but not as far as the southern and south-western Cape. The character which holds these forms together is the corona structure. The outer corona arises about halfway up the staminal column; the outer lobes are about equal in height to or overtop the staminal

Figure 57A. *Brachystelma circinatum:* a. plant with several annual branches from perennial stem, closely similar to forms described under the epithet *B. bolusii* from Graaff-Reinet, x ½; b. enlarged section of inflorescence, x 2 (*R. Bayliss* 2866 from Somerset East district, eastern Cape)

Figure 57B. *Brachystelma circinatum:* a. plant first named *B.galpinii,* x 1; b. flower with corolla lobes united at tips and showing corona within base, x 4; c. corona with outer lobes deeply bilobuled, erect, inner lobes linear incumbent, x 6; d. pollinia attached to small carrier, x 50 (adapted from Flower. Pl. Afr. 536)

Figure 57C. *Brachystelma circinatum:* a. form of this very complex group, from Pilgrim's Rest district above Strijdom Tunnel, grassveld in dolomite soil, x 2/3; b. flower showing unusual narrowing near base, x 2 (*D. Hardy* 4091); c. another form near railway line at Dullstroom, eastern Transvaal, x 1; d. flowers freshly opened, x 2 (*A. Vorster,* photo: Adela Romanowski)

Figure 57D. *Brachystelma circinatum:* a. plant from Rayton Ridge near Bloemfontein, tentatively left as a form of this very variable concept, with more slender corolla lobes and less developed outer corona lobes, x 1; b and c. flowers of different ages and stages of opening, x 2½ (*S.Rawlinson,* equal to specimens collected by *J.van Zyl* near Fauresmith, photo: Adela Romanowski)

column and are divided almost to their base into 2 lobules. As for the vegetative characters there seems endless variation in leaf shape and corolla and gradations occur between the extremes. I am unable to suggest any workable subdivision into subspecies or lower categories of classification. I was encouraged to take the rather drastic step of reducing several established names to synonymy by a report by A.A.Bullock of the Royal Botanic Gardens Kew staff some years ago. He said he was of the opinion that *B.pallidum, B. galpinii, B.ovatum, B.bolusii* and *B.commixtum* were not specifically distinct from *B.circinatum.* I added *B.filiforme, B.cinereum, B.undulatum, B.zeyheri* and *Dichaelia forcipata,* which makes a formidable list, but I omitted to repeat *Dichaelia forcipata* in Flora S.Afr. 27,4:33 (1980). A new approach to the problem with adequate living material might yield better results but this task must be left to a future worker.

On the other hand while examining the material of *B.circinatum* a few nonconforming but superficially similar specimens from the Griqualand West to northern OFS area, were segregated under the name *B.dimorphum* dealt with in the previous species. The difference is in the much reduced outer corona and hairiness of the inner surface of the corolla towards the base.

Also on the fringe of *B.circinatum* is a further complex first recorded by J.A.van Zyl, 1027, on the townlands of Fauresmith in the OFS and later by S.Rawlinson from near Bloemfontein. These have longer narrower leaves and long slender corolla-lobes together with shorter outer corona-lobes and generally very little pubescence. The classification has been left in abeyance because of the uncertainty of the best method of treatment, see Figure 57d.

58. **Brachystelma stenophyllum** *(Schltr.) R.A.Dyer* in Bothalia 10,2:376 (1971); in Flora S.Afr. 27,4:34 (1980) [Fig.58]

Perennial dwarf herb with tuber. *Tuber* 30-60 mm diameter, depressed. *Stem* 1, produced annually, sparsely branched above, up to about 150 mm tall. *Leaves* narrow, 20-50 mm long, 1,5-5 mm broad, slightly folded upwards, with midrib prominent on lower surface. *Flowers* 2-3 together, lateral at the nodes on stalks 3-8 mm long. *Corolla* 12-20 mm long, minutely hairy on outer surface; tube more or less narrowly bell-shaped, 5-8 mm deep, up to 7 mm diameter at mouth, maroon within; lobes 9-14 mm long, united at tips, very narrow with reflexed margins.

Brachystelma stenophyllum, a very slender leaved species, extends from northern SW Africa near Grootfontein-Kalahari sandveld of Botswana to the northern Transvaal and also to the northern area of the OFS near Odendaalsrus. It was first recorded in SW Africa by Dinter who suggested the genus *Ceropegia.* However Schlechter decided to place it in a separate genus with the name *Siphonostelma stenophyllum* but this seems unnecessary in view of

Figure 58. *Brachystelma stenophyllum:* plants from near Hammanskraal about 40 km north of Pretoria, *a.* in flower; *b.* in fruit, both about life size (*D.Hardy,* 2206)

the broad basis for *Brachystelma.* In the northern Transvaal the first collection was made by E.E.Galpin in the same period as Dinter. D.Hardy and others have corroborated this in more recent times. It is certainly a connecting link with *Ceropegia.*

59. **Brachystelma barberae** *Harv.* ex *Hook.f.* in Curtis's bot.Mag.:t.5607 (1866) as *barberiae;* R.A.Dyer in Flora S.Afr. 27,4:34 (1980) [Fig.59]

Perennial dwarf herb with tuber. *Tuber* 80-200 mm diameter depressed and becoming irregular in shape. *Stems* 1-few produced annually, up to about

Figure 59. *Brachystelma barberae:* a. plant from Transvaal, east of Pretoria, near Cullinan diamond mine, in sandy open grassveld, showing clusters of flowers with tips of corolla lobes united, x 1 (*E.A.Phillips* 118, cult. BRI, photo: H.King); b. plant of maximum size with tuber ±150 mm diameter and several annual stems each having produced clusters of flowers only two of which were fertilized under cult. BRI, x 1 (*Fanie Venter,* Rustenburg)

Plate 10. *Brachystelma huttonii* (species 35); in vicinity of type locality, Great Fish River catchment north of Grahamstown, eastern Cape, in karoid scrub (*R. Bayliss* 4404)

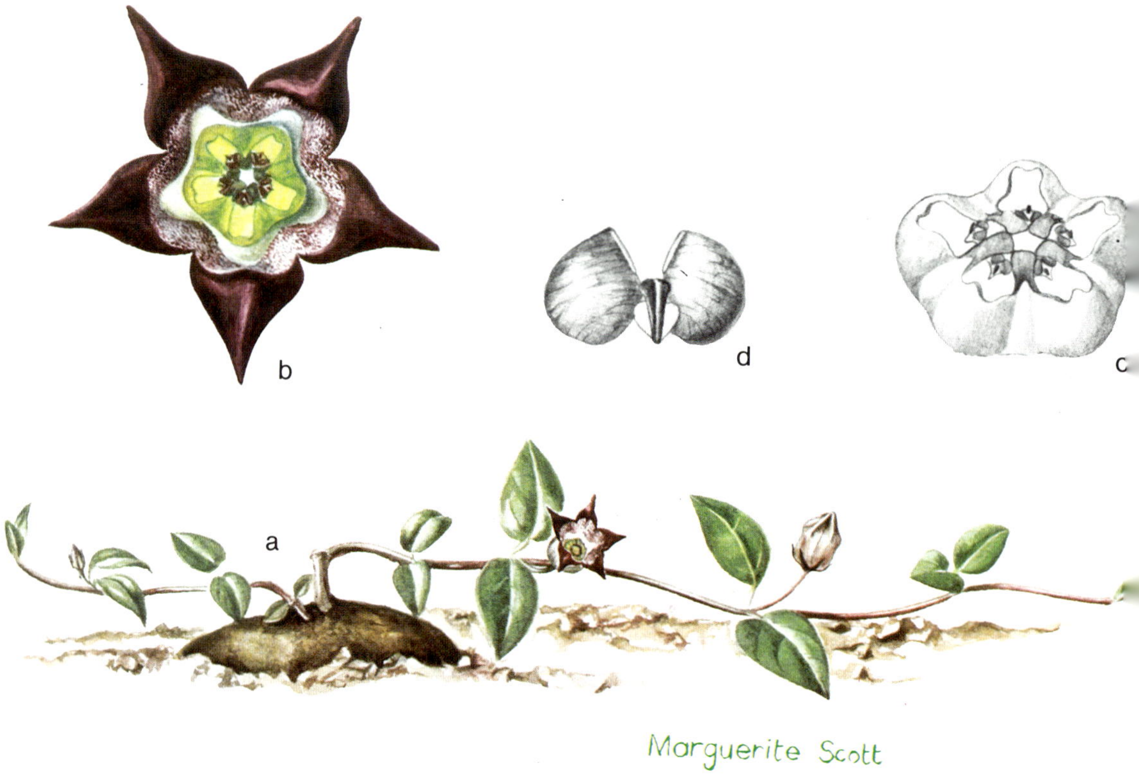
b
d
c
a
Marguerite Scott

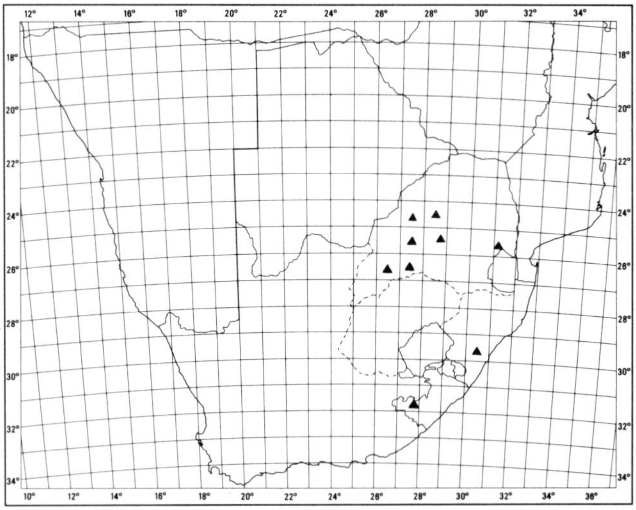

▲ *Brachystelma barberae* (59)

100 mm tall, somewhat coarsely hairy. *Leaves* up to about 100 mm long and 25 mm broad, coarsely hairy with most hairs on undersurface. *Flowers* sub-terminal and usually in two opposite sessile umbels of up to about 25 flowers in each, appearing, as one large terminal umbel. *Corolla* 20-45 mm long, divided nearly to base; tube shortly bell-shaped 5 mm deep, with crimson brown bands on a yellowish background within; lobes slender from near the base with recurved margins, united at tips into a cage-like structure; greenish outside, maroon to crimson-brown on inner face.

Brachystelma barberae, named after Mrs Barber (Miss Mary Bowker), is certainly one of the most spectacular species in the genus for the number, size and shape of the flowers. The illustration in Curtis's Botanical Magazine, plate 5607, was based on an original painting by Miss Bowker of a specimen collected in the Tsomo River valley in the Transkei, which had been collected by her brother, Mr Henry Bowker. In the text to the plate the tubers are said to have been eaten by the Transkeians. This is of particular interest since the edibility of the tubers is vouched for by the aborigines along its path of distribution into the northern Transvaal from where it extends further north into Zimbabwe. For this reason one might have expected its occurrence to have been reduced to rarity as in the case of many other species. Although of a scattered distribution over most of the area, records confirm that it is not infrequently exceedingly common locally. The size of the tubers, up to the maximum, must

Plate 11. *Brachystelma ngomense* (species 39); a. tuber giving rise to prostrate stems, tuber usually submerged below ground surface, bud broadest at the throat, x 1; b. flower face view, x 4; c. corona forming outer tube with slightly incurved margin and confluent with short inner lobes incumbent on base of filaments, x 8; d. pollinia with very short connectives to winged carrier, x 60 (*O.Hilliard & B.L.Burtt* 8441, Ngome, Zululand)

Figure 60. *Brachystelma natalense:* the typical form preserved in the Natal Herbarium, Durban, remarkable for the minute flowers on very slender thread-like stalks at the apex of the growing stem and the developing fruits on the previous season's growth on the much thickened stalks, x 2/3 (*J.Medley Wood* 176 = 410)

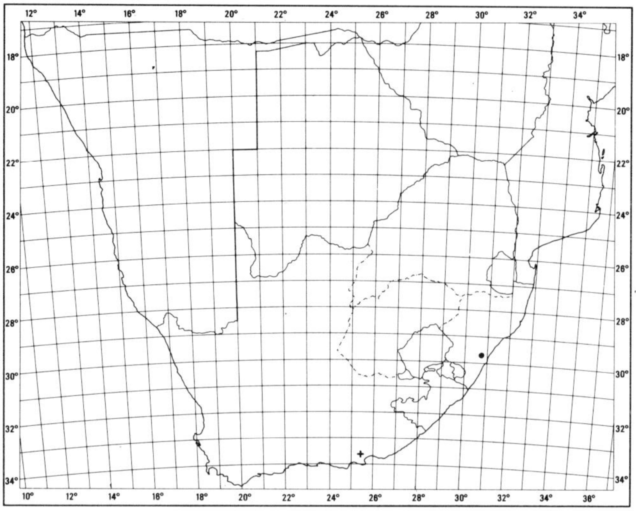

● *Brachystelma natalense* (60)
+ *Brachystelma comptum* (61)

be related to its age. The age of one of about 200 mm diameter is likely to be quite high and is also likely to be considerably tougher than the soft small ones more usual in the genus. Another feature of interest is that it conforms in that the flowers possess a most unpleasant odour and further it has a durable pollination mechanism which would encourage better than average seed production.

60. **Brachystelma natalense** *(Schltr.) N.E.Br.* in F.C. 4,1:850 (1908); R.A.Dyer in Flora S.Afr. 27,4:36 (1980) [Fig.60]

Perennial herb with tuft of somewhat succulent roots. *Stem* single, produced annually, up to 600 mm tall with rather coarse spreading hairs, leafy. *Leaves* rather broad, overlapping, 25-55 mm long, 16-40 mm broad. *Flowers* 2-4 lateral at the nodes towards apex of stem, small, on delicate stalks up to 20 mm long, becoming curved and robust on fertilization. *Corolla* 4-5 mm diameter, lobed to two-thirds; tube shallowly cup-shaped; lobes 2 mm long, spreading, greenish.

Brachystelma natalense was recorded by Medley Wood in 1883 from near Inanda, north of Durban in Natal at an altitude of approximately 600 m and since then nothing further has been added. Surely someone will wish to be the rediscoverer after a lapse of 100 years. The plant itself is remarkable for its robustness and the smallness of its flowers on thread-like stalks. Only after fertilization does the flower stalk elongate, thicken, and strengthen out of all proportion to its former self, supporting the comparatively large follicles (seed capsules) until after seed-dispersal late in the season.

Figure 61. *Brachystelma comptum:* a. plant from site overlooking Settlers' Dam near Grahamstown, showing flower with pinched-in waist, x 1; b. flower face view showing long clavate hairs from pinched-in waist and corona in centre, x 3/2 (*Northover* ex *Bayliss* 8163, cult. BRI, photo: Adela Romanowski)

61. **Brachystelma comptum** *N.E.Br.* in Fl.Cap. 4,1:854 (1908); R.A.Dyer in Flora S.Afr. 27,4:36 (1980) [Fig.61]

Perennial herb with rootstock producing a cluster of fleshy roots. *Stem* annual from rootstock, sparingly branched above, up to about 150 mm tall (more under cultivation) shortly hairy. *Leaves* shortly stalked, spreading, 10-160 mm long, 6-10 mm broad, thinly hairy on both sides or only on lower surface. *Flowers* solitary, lateral at the nodes. *Corolla* appearing to have a nearly flat base, inverted at pedicel attachment, constricted below the middle in bud, the lobes spreading from the constriction on opening, exposing maroon colour and with long clavate-vibratile hairs in the throat.

Brachystelma comptum, with the flower 'adorned with hair', has been collected only twice in 150 years. The early botanical collector Zeyher made the first discovery near Uitenhage in the Swartkops River Valley of the eastern

Cape. Recently Roy Bayliss was instrumental in following up a second record in Howiesonspoort near Grahamstown. The story started when Mr N.Northover noticed a strange looking little plant in a patch of veld which he was preparing for a new lawn. This happened at a time when one began to think that the species may have become extinct. Roy's intervention gives *B.comptum* a new lease of life. The species has quite a strong affinity with *Brachystelma gerrardii* from Natal and Transvaal but the coronal structures are distinctive as can be observed in the dissections published in the figures.

62. **Brachystelma gerrardii** *Harv.* in Thes.Cap. 2:61, t.196 (1863); R.A.Dyer in Flora S.Afr. 27,4:38 (1980) [Fig.62]

Perennial herb with rootstock producing a cluster of fleshy roots. *Stem* single annually, sparingly branched up to about 350 mm tall with long hairs. *Leaves* shortly stalked, up to about 40 mm long and half as broad. *Flowers* solitary, lateral at the upper nodes, on slender stalks 15-35 mm. *Corolla* with constric-

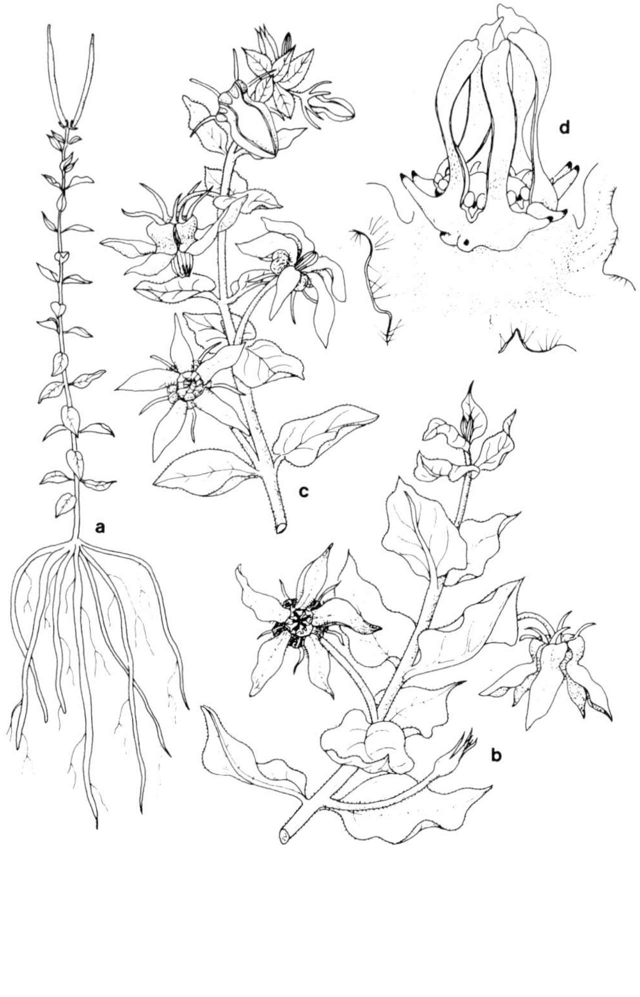

Figure 62. *Brachystelma gerrardii:* a, habit with cluster of fleshy roots, x 1/8; b and c, flowering terminal branches with bud and open flower showing the characteristic constriction above the base round the corona, x ½; d, corona arising from base of corolla, outer lobes spreading, bifid, inner lobes connivent-erect, up to 6 mm long, x 3½ (after Flower. Pl. Afr. 1686).

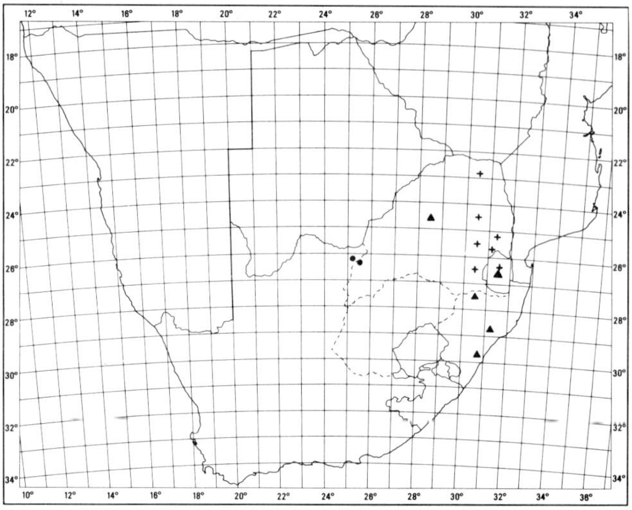

▲ *Brachystelma gerrardii* (62)
\+ *Brachystelma macropetalum* (63)
● *Brachystelma canum* (64)

tion below middle in bud, divided almost to base; base inverted; lobes contiguous round staminal column, then with slightly pinched-in margin gradually spreading, up to 20 mm long, 4-6 mm broad, inner surface minutely papillate, dark green to metallic blue-green or black, greenish-grey or yellowish within base, margins ciliate with long, simple or clavate, reddish-purple, vibratile hairs.

Brachystelma gerrardii named after the botanical collector W.T.Gerrard by William Harvey in 1859 in the following terms: 'as one of the most curious of the many curious Asclepiads which we owe to his researchers'. The species has a scattered distribution from the midlands of Natal to Swaziland and northern Transvaal but is usually rare or very rare within this range. When Kathleen Gordon-Gray wrote the text for a plate 1686 in Flowering Plants of Africa, 1974, she commented on the remarkable colour of the flowers pointing out that in southern Natal the colour is bright metallic green while to the north it is dark green approaching black, velvety above. She explained that the velvety texture results from closely packed papillae, which, when they lose turgor on drying, collapse away from one another to produce a misleading, hairy appearance in dried specimens. The long reddish-purple, vibratile hairs along the margin of the lobes towards the base are another noteworthy feature, found also in *B.comptum*, its nearest relative.

63. **Brachystelma macropetalum** *(Schltr.) N.E.Br.* in Fl.Cap. 4,1:852 (1908); R.A.Dyer in Flora S.Afr. 27,4:39 (1980) [Fig.63]

Perennial herb with rootstock producing a cluster of fleshy roots. *Stem* single, produced annually, simple or occasionally sparingly branched, up to 600 mm

Figure 63. left: *Brachystelma macropetalum,* herbarium sheet from Transvaal Museum now in BRI, coll. F.A.Rogers near Sabie, reduced. right: *Brachystelma canum,* type sheet in BRI, reduced.

tall, softly hairy. *Leaves* shortly stalked, ascending, up to about 35 mm long and two-thirds as broad, softly hairy. *Flowers* solitary lateral at the nodes on stalks up to about 25 mm long and with conspicuous sepals up to about 12,5 mm long. *Corolla* divided almost to base, the united portion flattish or saucer-shaped; lobes narrow from broader base spreading yellow to greenish, marked with reddish to brownish dots.

Brachystelma macropetalum with comparatively 'long' petals, is apparently frequent to common in damp situations on the high mountains of the eastern escarpment (Transvaal Drakensberg) with outliers further north on the Zoutpansberg. One's imagination could lead to the suggestion that this species resembles a large hairy spider and that the corona with dark violet-brown to black inner lobes represents the fearsome head and mandibles. Fortunately a living plant collected by D.Hardy in 1982 has become available for illustration in colour at BRI.

64. **Brachystelma canum** *R.A.Dyer* in Bothalia 12,2:254 (1977); in Flora S. Afr. 27,4:39 (1980) [Fig.63]

Perennial herb with rootstock (presumed) producing a tuft of fleshy roots. *Stem* single, produced annually, unbranched or sparsely branched from near base. *Leaves* nearly stalkless, 20-40 mm long, 5-10 mm broad, narrowed gradually to base, finely grey-hairy, margins slightly undulate and folded upwards. *Flowers* about 2, lateral at the upper nodes, opening in succession. *Corolla* 9-11 mm long, lobed almost to base, united for 1,5 mm; lobes narrow, with margins recurved, finely hairy on outer surface.

Brachystelma canum, covered with 'grey' hairs is based on a single specimen collected by John Acocks in February 1956. He collected it 25 km north-east of Setlagoli on the road to Mafeking in sandy Terminalia-veld. Normally one would not describe a species on such meagre material, but in this case I hope it will encourage collectors to make special efforts to discover adequate study material. It is unlike John Acocks not to have recorded the root system which suggests that to unearth it presented unusual difficulties. The illustration gives an indication of what to look for in the veld.

65. **Brachystelma schizoglossoides** *(Schltr.) N.E.Br.* in Fl.Cap. 4,1:849 (1908); R.A.Dyer in Flora S.Afr. 27,4:39 (1980) [Fig.64.1]

Perennial herb with rootstock producing a cluster of fleshy roots. *Stem* single, simple, produced annually, up to about 250 mm tall, slender. *Leaves* slender, stalkless, up to about 45 mm long. *Flowers* 1-3 together lateral at the nodes,

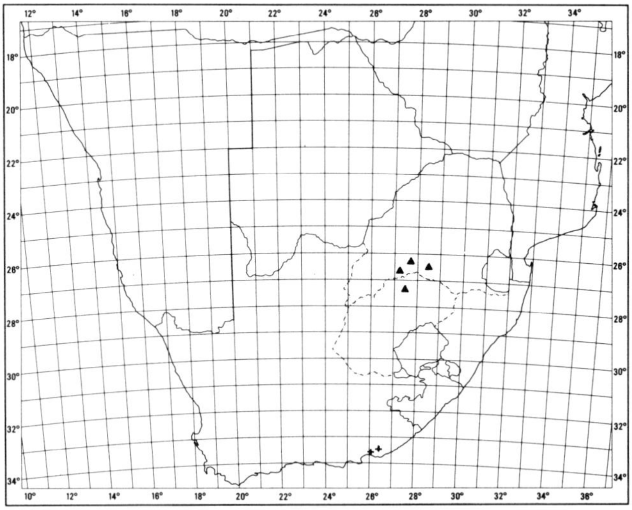

\+ *Brachystelma schizoglossoides* (65)
▲ *Brachystelma ramosissimum* (66)

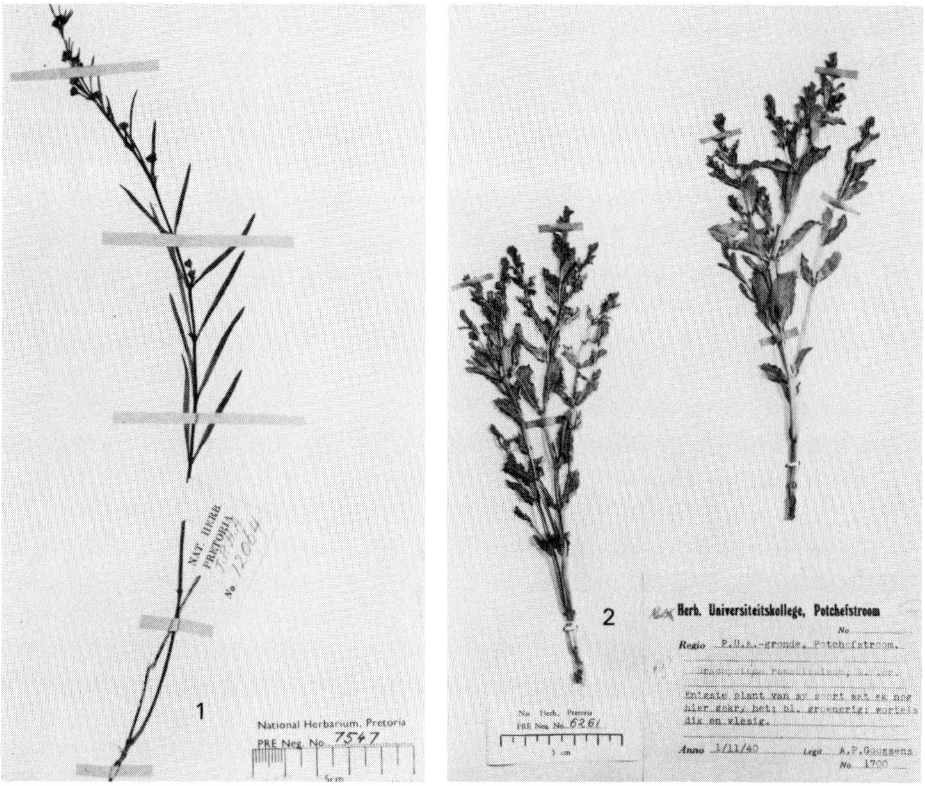

Figure 64. 1. *Brachystelma schizoglossoides,* specimen in PRE from near type locality, insignificant flowers, x ⅓ (*J.Acocks* 12064, Bathurst, eastern Cape). 2. *Brachystelma ramosissimum* specimen from near type locality, broader leaves than usual, small flowers, x ¼ (*A.P.Goossens* 1700, Potchefstroom, Transvaal).

rarely axillary, shortly stalked. *Corolla* 2-2,5 mm long, lobed up to two-thirds, green and inconspicuous; tube saucer-shaped much shorter than staminal column; lobes slightly spreading, with incurved tips.

Brachystelma schizoglossoides is restricted to the grassveld of the Albany and Bathurst districts of the eastern Cape according to present records. It has even less to recommend it for cultivation than most other species and is unlikely to attract the attention of a non-botanist.

66. **Brachystelma ramosissimum** *(Schltr.) N.E.Br.* in Fl.Cap. 4,1:855 (1908); R.A.Dyer in Flora S.Afr. 27,4:40 (1980) [Fig.64.2]

Perennial herb with rootstock producing tuft of long fleshy roots. *Stem* single (2), produced annually, up to about 150 mm tall, richly branched with the branches angular or ribbed. *Leaves* with little or no stalk, narrow, 20-30 mm long, 2-5 mm broad. *Flowers* 2-4 together, lateral at the nodes developed in succession. *Corolla* 3-4 mm long, divided nearly to base, white to pale yellow or greenish; lobes small with recurved margins, not greatly spreading.

Brachystelma ramosissimum is a dainty 'much branched', free flowering, little herb but not one for the collector to enthuse over greatly. It is not uncommon in the western Transvaal grassveld near Potchefstroom extending to near Johannesburg and southwards over the OFS border.

67. **Brachystelma sandersonii** *(Oliv.) N.E.Br.* in Fl.Cap. 4,1:850 (1908); R.A. Dyer in Flora S.Afr. 27,4:40 (1980) [Fig.65]

Perennial herb with rootstock producing a cluster of long fleshy roots. *Stem* single (2), unbranched or sparingly branched about the middle, up to about 450 mm tall. *Leaves* without or with very short stalk, ascending, mainly narrow; 20-40 mm long, 3-12 mm broad. *Flowers* (1)2-several together, lateral at the nodes, developing in succession but sometimes more than one open at a time. *Corolla* up to 4 mm long, divided almost to base; lobes not greatly expanding, with recurved margins, keeled down inner surface.

Brachystelma sandersonii named after John Sanderson, journalist and naturalist with artistic ability whose headquarters were in Durban during the middle 19th century. The species is rather featureless with a distribution in coastal Natal extending into Zululand and with an outlier record in the eastern Cape near King William's Town. Some authors have in the distant past excluded it

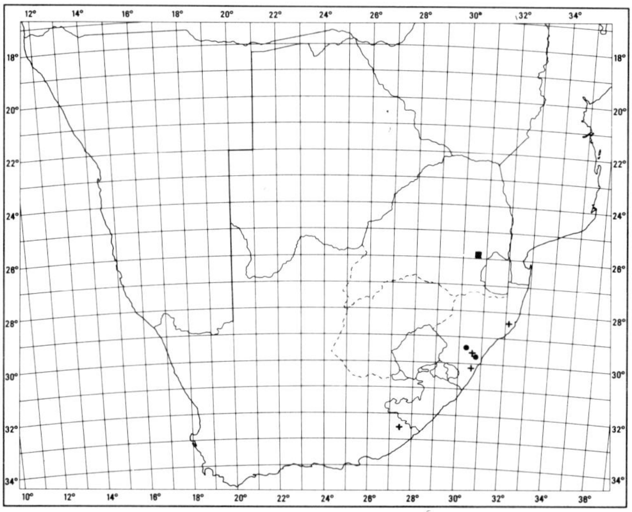

+ *Brachystelma sandersonii* (67)
● *Brachystelma franksiae* (68)
■ *Brachystelma longifolium* (69)

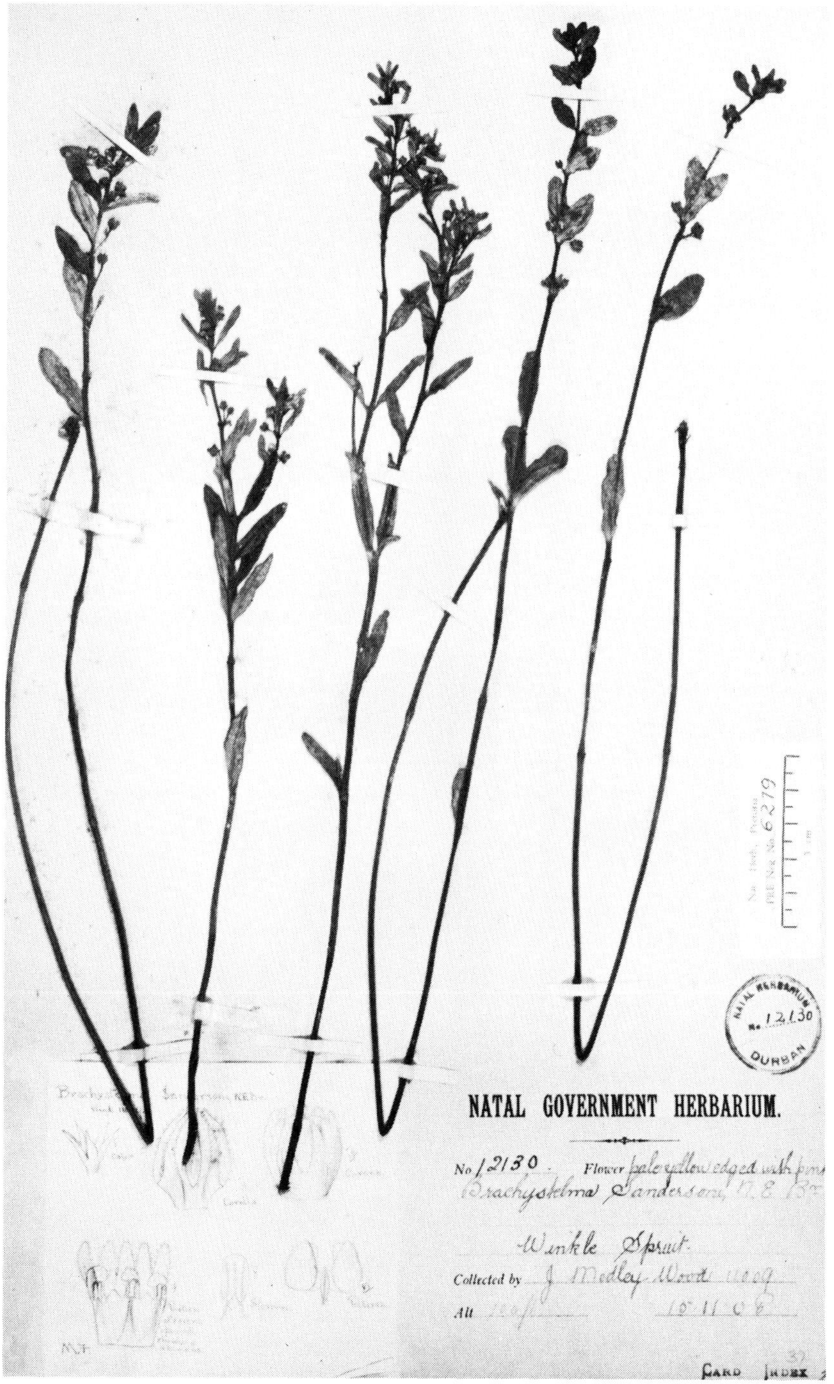

Figure 65. *Brachystelma sandersonii,* specimen in Natal Herbarium, Durban, from near the type locality (*Medley Wood* 11009, Winklespruit, Natal).

and its close allies from the genus *Brachystelma* because of the root-system and flowers and placed them either under the names *Brachystelmaria* or *Lasiostelma,* but to follow this lead would have far-reaching implications and yield very doubtful benefits.

68. **Brachystelma franksiae** *N.E.Br.* in Medley Wood, Natal Plants 6:t.588 (1912); R.A.Dyer in Flora S.Afr. 27,4:40 (1980) [Fig.66]

Perennial herb with rootstock producing a cluster of long, fleshy roots. *Stems* 1-2, produced annually, erect, up to 300 mm tall, sparingly branched. *Leaves*

Figure 66. *Brachystelma franksiae:* a, cluster of fleshy roots, x ½; b, flowering stem sparsely branched, x ½; c, flower, x 1½; d, corolla-lobe with papillae-like hairs on margin, x 3; e, corona, x 5; f, pollinia, x 15 (after *Medley Wood,* Natal Plants, 6:t.588).

without or with short stalk, usually rather narrow, up to 50 mm long. *Flowers* 1-3 together lateral at the nodes, developed successively but sometimes more than one open at a time. *Corolla* divided nearly to base, white tinged with pink; tube saucer-shaped; lobes 6-7 mm long, erect, with recurved margins and keeled on inner face.

Brachystelma franksiae is found mainly in the midland grassveld of Natal,

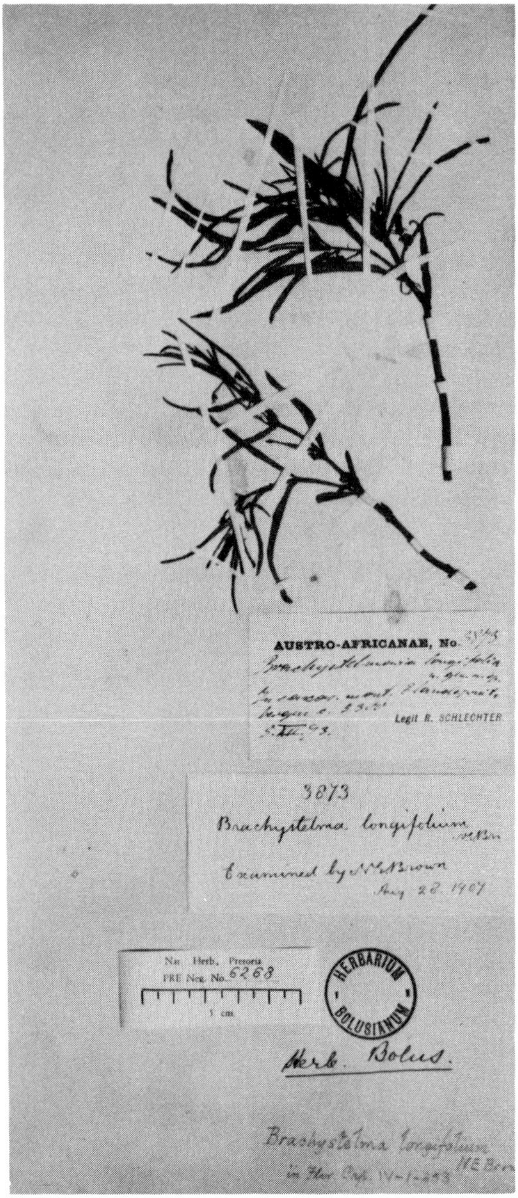

Figure 67. *Brachystelma longifolium*, lectotype specimen in Bolus Herbarium (*Schlechter* 3873, Elandspruit Mt, eastern Transvaal).

being fairly frequent from Camperdown to Pietermaritzburg. In describing the species N.E.Brown said he had very much pleasure in dedicating this interesting species to its discoverer, Miss Millicent Franks who had so ably assisted Medley Wood in his labours, and to whom we are indebted for many admirable drawings published in the series on Natal Plants. Interesting the species is, but not likely to become a collector's favourite.

69. **Brachystelma longifolium** *(Schltr.) N.E.Br.* in Fl.Cap. 4,1:853 (1908); R.A.Dyer in Flora S.Afr. 27,4:41 (1980) [Fig.67]

Perennial herb with rootstock producing a cluster of long fleshy roots. *Stem* 1-2, produced annually, 100-300 mm tall, sparsely branched. *Leaves* without or with very short stalk, narrow, 30-70 mm long, up to 4 mm broad. *Flowers* 1-2, lateral at nodes. *Corolla* about 10 mm long, divided nearly to base, possibly greenish tinted brown; lobes narrow, with margins reflexed, with minute papilla-like hairs or clavate purple hairs to halfway.

Brachystelma longifolium with comparatively long slender leaves was first recorded by Schlechter in 1893 in a stony place on the Elandspruit mountains in the eastern Transvaal. It has not been collected again since. To rediscover it and amplify our meagre knowledge of the species would certainly be a useful achievement. As with other related species *Schlechter* first described it under the generic name *Brachystelmaria* and then transferred it to *Lasiostelma.* It was N.E.Brown at the Royal Botanic Gardens, Kew, who decided on its inclusion, with its several relatives, in the umbrella genus *Brachystelma.*

Ceropegia

Ceropegia *L.*, Sp.Pl.1:211 (1753); R.A.Dyer in Flora S.Afr. 27,4:43 (1980); Bruyns, Ceropegias of the Pretoria District in Aloe 18,1 & 2:21 (1980) [Fig.68]

Perennial herbs, sometimes succulent, with tubers or fusiform or occasionally fibrous roots. *Stems* dwarf to elongate, prostrate to erect, sometimes dangling, scrambling or twining. *Leaves* opposite, stalked or stalkless, sometimes obsolete on fleshy stems. *Flowers* in umbel-like clusters, racemes or 1-several developing together or in succession, mainly lateral at the nodes. *Corolla* with a tube longer than the lobes, often somewhat inflated near base; with or without hairs on outside often with hairs within; lobes recurved or pendulous or connate at tips into a cage-like form or forming an umbrella-like canopy, or extending into a capitate and winged apical structure, with margins often replicate, often ciliate. *Fruits* consisting of 1 or usually two elongate hornlike structures, producing winged seeds crowned with a tuft of fine hairs.

Ceropegia based on *C. candelabrum* from Sri Lanka (Ceylon) and India is represented by over 160 species with a range from the Canary Islands in the west, throughout Africa into India and the Far East to New Guinea and northern Australia.

At the specific level all previous workers, including the authorities Schlechter, Schumann, N.E.Brown and Huber have remarked on the difficulty in accurately defining inter-related species. Huber was the last to monograph the genus *Ceropegia* in Mems Soc. broteriana in 1958. The work has been used extensively in my work although I have often deviated from his conclusions.

The most noteworthy feature about his work is that he dealt with the genus as a whole. He paid particular attention to published descriptions and illustrations in order to arrive at his conclusions on specific limits and synonymy. He found, as had Schumann, Schlechter, N.E.Brown and others before him, that certain morphological characters may exhibit considerable variability, for example, leaf-shape, whereas others, such as pubescence, may be relatively constant. Huber groups the species of the world into 21 sections, 19 of which are of his own making. His emphasis on pubescence in his grouping seems eventually to lead to some strange bedfellows. His field experience was nil.

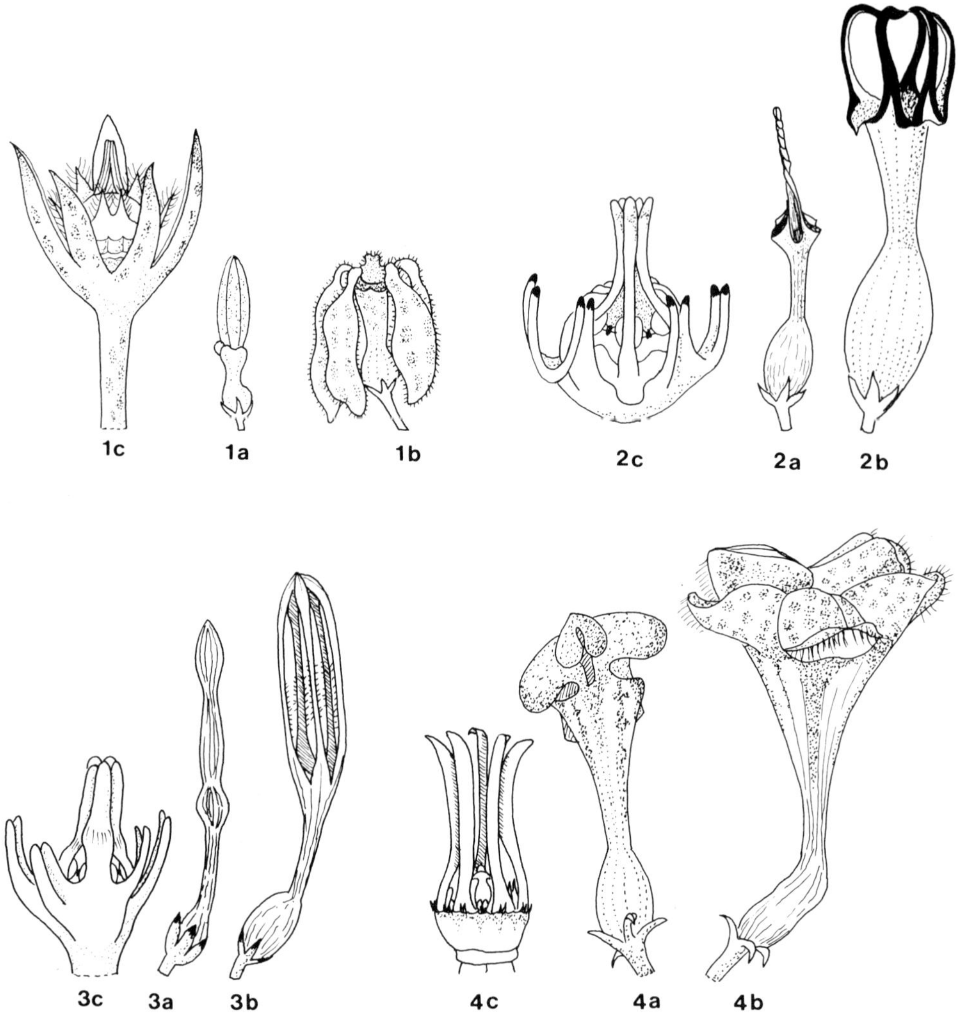

Figure 68. 1. *Ceropegia bowkeri* subsp. *bowkeri:* 1a, bud, x ¾; 1b, open corolla with pendulous lobes, x ¾; 1c, flower with corolla removed, showing calyx outside and corona within, x 4½. 2. *C.filiformis:* 2a, bud, x ¾; 2b, open corolla with cage-like (or crown-like) apex, x ¾; 2c, corona, x 7½. 3. *C.zeyheri:* 3a, bud, x ¾; 3b, open corolla with lobes forming elongated cage and hairs about middle of lobes, x ¾; 3c. corona, x 7½. 4. *C.sandersonii:* 4a, bud, x ¾; 4b, open corolla with fringed canopy, x ¾; 4c, corona with basin-shaped base and slender, erect inner lobes, x 7½ (adapted from PRE records).

With the distribution of the genus extending from China and the East Indies to the Canary Islands in the West and the Cape Peninsula in the South and the tendency of taxonomists to work on the floras of limited areas in isolation, it is not surprising that Huber was beset with a problem of dense synonymy. The final result was a synonymy of about two-thirds of the total number of published specific epithets.

In the circumstances it is not surprising either that one finds grounds for agreement and disagreement with Huber's conclusions. One matter of disagreement is the sinking of *Riocreuxia* in *Ceropegia.* It is here restored to generic rank. In cases of a species distribution extending from tropical Africa into southern Africa, Huber's nomenclature has been mostly accepted but not in cases where the evidence appeared to be inconclusive.

Huber, l.c., 112, 113, postulates that garden hybrids between *C.barklyi* and *C.rendallii* and between *C.barklyi* and *C.woodii (= linearis* ssp. *woodii*) have taken place in Europe. Several cases of suspected hybridization both in cultivation and in the wild have been recorded locally. I have found some inconstancy in my distinction between *C.barklyi* and *C.africana* in herbarium specimens, and in the limits of *C.cancellata, C.linearis, C.woodii* and other related species. The problems of relationship are most perplexing. Hybridization may well be implicated where there have been overlaps in the distribution ranges. In this respect the notes of Brownlee referred to under *C.cancellata* in the present account are of particular relevance.

One may imagine that the process of classifications would be simplified if one had a comprehensive large collection of living plants to study. In practice this sometimes multiplies one's problems because of the varying responses to the changed environment. Dwarf plants may elongate and twine, similar plants from the same environment in the wild may soon accentuate their differences in vigour. One is persuaded largely to neglect vegetative characters such as size and shape of leaves and to afford greatest importance in classification to the floral structure. But even in this respect there is not absolute uniformity in wild populations and the mental churning and assessment of characters continues. Bruce Bayer, at the Karoo Garden, Worcester, has grown seed from the wild with surprisingly inconsistent results. Seedlings from the same seed pod have been either hairy or not and flowers have shown variation in the colour patterns. In other cases flowers have shown structural variations, the reason for which is unexplained. Obviously there is a very wide field for further research.

KEY TO FOUR MAIN GROUPS OF *CEROPEGIA*

1 Rootstock producing a cluster of fusiform, fleshy or fibrous roots:
  2 Corolla-lobes free at apex (see *C.ampliata*) . . . . – **Group A** (1-7)
  2 Corolla-lobes fused at apex . . . . – **Group B** (8-26)
1 Rootstock producing a solid tuber, sometimes climbers with adventitious tubers along stem:
  3 Corolla-lobes free at apex . . . . – **Group C** (27-30)
  3 Corolla-lobes fused at apex . . . . – **Group D** (31-51)

Group A. Rootstock producing a cluster of fleshy or fibrous roots; corolla-lobes free at apex:

1 Herbs, erect, rarely up to 300 mm tall, usually in grassveld, leaves usually slender, herbaceous or slightly fleshy:
  2 Corolla-lobes pendulous, swaying in the breeze, oblong, narrowed into a short stalk (hinge) at base:
    3 Basal inflation of corolla-lobe subglobose, passing more or less abruptly into upper cylindric portion . . . . 1(a). **C.bowkeri** subsp. **bowkeri**
    3 Basal inflation of corolla-tube elongate, ovoid, passing gradually into upper portion . . . . 1(b). **C.bowkeri** subsp. **sororia**
  2 Corolla-lobes spreading gradually from tube, not hinged, linear to filiform:
    4 Corolla-lobes with a tuft of longish red hairs slightly above base; leaves linear:
      5 Corolla smooth on outside of inflated base; lobes slightly thickened upwards . 2. **C.tomentosa**
      5 Corolla papillate-scabrous on outside of inflated base; lobes evenly linear . . 3. **C.scabriflora**
    4 Corolla-lobes linear, replicate, pubescent on inner face but without tuft of hairs near base; leaves oblong or elliptic-oblong . . . . . 4. **C.rudatisii**
1 Succulent or semisucculent scramblers or climbers, leafless or with linear leaves, in scrubveld:
  6 Leaves rudimentary; succulent climbers:
    7 Corolla up to 70 mm long, nearly half divided; lobes usually spreading-recurved . 5. **C.stapeliiformis**
    7 Corolla up to about 50 mm long, about half divided; lobes opening star-like over the tube . . . . 6. **C.cimiciodora**
  6 Leaves linear; corolla-lobes twisted in bud with knob-like apex, becoming free and straight on opening and bearded with long hairs . . . . 7. **C.barbata**

Group B. Rootstock producing a cluster of fleshy or fibrous roots; corolla-lobes fused at apex:

1 Corolla-lobes twisted in bud, becoming untwisted on opening; corolla with basal inflation oblong-elliptic; leaves linear to filiform . . . . . . 8. **C.filiformis**
1 Corolla-lobes not twisted in bud:
  2 Leaves absent or much reduced and scale-like; stems succulent:
    3 Corolla with basal inflation simple, without secondary constriction:
      4 Corolla-lobes narrowed evenly to the united apex; corolla-tube ample, remaining more than 10 mm diameter above basal inflation; greenish-white, finely longitudinally veined . . . . 9. **C.ampliata**
      4 Corolla-lobes broadened towards apex, sometimes united into a canopy:
        5 Corolla-lobes with free portion or struts ±5 mm high, rarely up to 7 mm, united above into a flattish or dome-shaped canopy, with long, purple hairs, which are pendulous or directed inwards:
          6 Corolla-canopy ±1,5 mm diameter, dome-shaped with long clavate, vibratile hairs pendulous from the canopy margin . . . . 10. **C.fimbriata**

6 Corolla-canopy 10-12 mm diameter, becoming flattish with long, clavate hairs from the struts directed inwards; margins of struts incurved 11. **C.connivens**

5 Corolla-lobes with free portion 10-45 mm long, variably united, with or without long purple, clavate hairs:

7 Corolla-lobes 10-20 mm long, usually more or less broadly spathulate, kneed about the middle or towards base, sometimes united into an incised canopy, usually with long, irregularly spreading, clavate hairs . 12. **C.geniculata**

7 Corolla-lobes 20-45 mm long, narrowly spathulate towards apex, not kneed, with or without small patch of clavate, vibratile hairs from inner surface, often puberulous outside for short distance . . . . 13. **C.zeyheri**

3 Corolla with basal inflation with a secondary median constriction; corolla up to 60 mm long; lobes with long, clavate, vibratile hairs mainly in upper half . . . 20. **C.arenaria**

2 Leaves developed:

8 Corolla-lobes united above into a canopy either centrally depressed or somewhat conical, often with a fringe of hairs from the margin . . 14. **C.sandersonii**

8 Corolla-lobes not united into a canopy:

9 Corolla-lobes extended above the apex of fusion into a slender, long or short column terminated by an angled head or cage . . . 15. **C.distincta**

9 Corolla-lobes not extended above the apex of fusion:

10 Outer corona-lobes deeply bilobuled or corona sometimes appearing 10-toothed, usually with few long hairs; inner corona-lobes ± linear, incumbent-erect, usually with somewhat spreading tips:

11 Corolla with basal inflation ± globular, less than twice diameter of upper tube:

12 Outer corona-lobes bilobuled; corolla 25-35 mm long, lobes strongly replicate, forming conical cage, sometimes apiculate; leaves mostly ovate-elliptic, apiculate . . . . 16. **C.lugardiae**

12 Outer corona appearing 10-lobuled or -toothed, corolla ±20 mm long:

13 Corolla-lobes, one-third to one-half length of corolla, cage slender; leaves up to 70 mm long, 40 mm broad . 17. **C.purpurascens**

13 Corolla-lobes, ±quarter length of corolla, the cage short and broadly rounded; leaves rarely up to 30 mm . . 24. **C.carnosa**

11 Corolla with basal inflation oblong-elliptic, longer than broad, passing gradually into short waist of tube; corolla-lobes with infolded margins and long, clavate-vibratile, caducous hairs from lower half . . . . . 18. **C.cycniflora**

10 Outer corona-lobes forming five small pockets, with outer margin subentire or spreading chute-like or scoop-like or deeply and narrowly V-shaped (falsely appearing as subquadrate lobes behind inner lobes):

14 Corolla with basal inflation with median constriction usually corresponding with internal collar:

15 Corolla up to 50 mm long; lobes rarely more than 15 mm long, without long clavate hairs; outer corona forming 5 pockets with entire outer margin; stems climbing and twining . . 19. **C.nilotica**

15 Corolla up to 60 mm long; lobes usually ±25 mm long, linear with spathulate apex, with long, clavate, spreading hairs near apex; stems sometimes prostrate without support . . . 20. **C.arenaria**

14 Corolla with basal inflation without median constriction and sometimes reduced to an indistinct swelling:
16 Corolla with basal inflation sausage-shaped, passing gradually into upper tube:
17 Stems climbing, diffuse; corolla ±25 mm long, ± half divided; tube slender with abruptly spreading sinuses; lobes linear to filiform . 21. **C.stenantha**
17 Stems prostrate; corolla 60-80 mm long, ± one-third divided:
18 Corolla-lobes forming an elongate pyramid 25 mm long to the united apex . . . . 22a. **C.radicans** subsp. **radicans**
18 Corolla-lobes cordate-oblong, forming an obtuse, cage-like structure about 15 mm long, slightly narrowed about the middle . . 22b. **C.radicans** subsp. **smithii**
16 Corolla with basal inflation more or less globose or almost obsolete:
19 Outer corona-lobes spreading chute- or scoop-like:
20 Succulent, glabrous climber and twiner; leaves soon decreasing in size on annual growth; flowers in dense cymes; corolla 25-50 mm long, ± one-third divided . . 23. **C.crassifolia**
20 Herbaceous, slender, puberulous climber and twiner; leaves ovate-cordate, ±25 mm long; corona united at base with inner lobes arising from within basin . . . . . 26. **C.dubia**
19 Outer corona-lobes forming five shallow pockets, with outer margin deeply V-shaped and adjacent halves of 'lobes' forming acute or squared shoulders confluent with the inner corona-lobes; corolla 20-25 mm long, about one-third divided into a somewhat rounded cage, basal inflation almost obsolete . . 25. **C.racemosa** subsp. **setifera**

Group C. Rootstock producing a solid tuber; corolla-lobes free at apex; dwarf plants (climbing in *C.multiflora*):

1 Dwarf herbs, rarely up to 0,3 m tall, in grassveld or open bushveld:
2 Corolla variably curved, drooping, with triangular lobes much shorter than tube, dark brown to blackish, puberulous; corona with deep cylindric outer tube 3 mm long, with small inner lobes arising from near base within . . . 27. **C.pygmaea**
2 Corolla with lobes ± length of tube:
3 Herb, dwarf with 1-few stems, rarely over 50 mm high, with few – many-flowered fascicles of flowers opening together; corolla dark maroon to blackish, 10-12 mm long, glabrous, divided to slightly more than halfway; lobes obovate-oblong, replicate, inflexed, acute at apex . . . . 28. **C.mafekingensis**
3 Herbs, somewhat elongate, 150-300 mm tall with spaced leaves; flowers single (2); corolla 75-100 mm long; leaves linear to filiform, sometimes with thickened tips, with red or purple hairs near base:
4 Corolla-lobes ±47 mm long, with tuft of red clavate hairs above base, outer corona-lobes bilobuled; inner incumbent-erect . . . 29. **C.dinteri**
4 Corolla-lobes ±50 mm long, with expanded tips and purple hairs near base;

outer corona-lobes forming five pockets with outer margin laterally toothed; inner incumbent-erect, reflexed . . . . 30. **C.antennifera**

1 Climbing and twining; corolla-lobes filiform or hair-like . . . 35b. **C.multiflora** subsp. **tentaculata**

Group D. Rootstock producing a solid tuber; stem dwarf or elongate; corolla-lobes fused at apex, sometimes into a cap or small canopy:

1 Erect or suberect dwarf herbs (in *C.conrathii* sometimes slightly fleshy), mainly in grass-veld:

2 Leaves linear-elongate; stems usually single from tuber:

3 Corolla 100-120 mm long, with lobes twisted together in upper half; flowers few, solitary at nodes . . . . 31. **C.insignis**

3 Corolla 45-70 mm long; lobes straight:

4 Corolla-lobes very slender, much longer than tube; tube bent at right angles . 32. **C.stentiae**

4 Corolla-lobes far shorter than tube, flat-topped; lobes with long clavate-vibratile hairs . . . . 33. **C.turricula**

2 Leaves ovate-oblong to oblanceolate, somewhat fleshy; sometimes inflorescence produced before normal leaves; stem often few–many from tuber; flowers 1 – few together; corolla 20-35 mm long:

5 Corona shallowly basin-shaped at base, with outer margin deeply U-shaped and confluent with inner lobes; corolla-lobes linear from very near base, glabrous, forming rounded cage . . . . 34. **C.conrathii**

5 Corona deeply campanulate, margin with five small and five larger undulations in rim; inner corona-lobes arising from near base of outer basin; corolla-lobes linear from ovate-lanceolate base, with few hairs from inner margin, forming oblong cage 45. **C.fortuita**

1 Climbing, scrambling or prostrate, often partly succulent herbs:

6 Basal corolla-inflation ± globose, shorter than the upper cylindric or narrowly funnel-shaped tube; stems and leaves mainly glabrous (except *C.pachystelma*):

7 Inflorescences at nodes ± subsessile, usually many-flowered with two or more flowers open at same time; buds tapering to a fine point:

8 Corolla-lobes linear from very near base, to tips forming a rounded cage. corona shallowly saucer-shaped, five-lipped, inner lobes spathulate . . . . 34(a). **C.floribunda**

8 Corolla-lobes hair-like from about middle, becoming almost horizontal over the corolla-throat; corona cupular 1 mm deep, inner lobes linear or spathulate . 35(a). **C.multiflora** subsp. **multiflora**

7 Inflorescences usually pedunculate, few-flowered, rarely more than one flower of an inflorescence open at the same time:

9 Corona mainly with campanulate or saucer-shaped base, sometimes fluted or lipped but not extended into distinct lobes:

10 Corolla glabrous on outer surface:

11 Inner corona-lobes, laterally compressed, broadly falcate, spreading, obtuse:

12 Corolla-lobes united at apex into an umbrella-like canopy ±10 mm diameter, supported by slender struts . . . . 37. **C.rendallii**

12 Corolla-lobes united at tips forming an open-cage-like structure:
13 Corolla-lobes about equal to or longer than slender tube, forming a long slender beak in bud, often twisted; habit variable . . . . 38. **C.barklyi**
13 Corolla-lobes one-third to one-half length of corolla, forming slender beak; tube somewhat trumpet-shaped at mouth . . . . 39. **C.africana**
11 Inner corona-lobes linear, lanceolate, spathulate, sometimes recurved and appearing narrowly falcate in side view:
15 Corolla-lobes broadly spathulate, with the united tips forming a small cap or canopy:
16 Corolla-tube somewhat funnel-shaped, not papillate within base; corona with five shallow pockets, inner corona-lobes ± linear-loblong, not or only very slightly narrowed below middle . . 40. **C.occulta**
16 Corolla-tube cylindric, papillate within base, very slightly expanded to mouth; corona with outer tube overtopping staminal column, inner corona-lobes ± spathulate, narrowed to a slender base; leaves often shed before flowering 41(a). **C.decidua** subsp. **decidua**
15 Corolla-lobes narrowly spathulate, linear or tapering to apex:
17 Corolla-tube expanded to a trumpet or funnel-shaped mouth:
18 Plants ± upright in nature with support of shrublets, ±300 mm high; leaves fleshy, subcylindric-oblong; corona united at base, ±1,5 mm deep with suberect, linear lobes from the inner margin . . . . 42. **C.occidentalis**
18 Plants climbing and twining or prostrate or hanging, sometimes dwarf; basal leaves flattened:
19 Corolla-lobes ± length of tube, glabrous or rarely with short hairs, narrowed in bud gradually from base to apex without constriction; corona subcampanulate with five-fluted tube; inner lobes erect-recurving, acuminate, grooved down face, angled down back . 43. **C.cancellata**
19 Corolla-lobes usually much shorter than tube, with long or short hairs from inner surface; inner corona lobes dorsally flattened, linear or spathulate:
20 Corolla-lobes spathulate, densely or thinly bearded; corona shortly united at base and confluent with base of ascending-recurving subacute inner lobes . . . . 44. **C.woodii**
20 Corolla-lobes narrowed from base to apex, shortly pubescent from inner surface; corona with campanulate outer tube from within which arise incumbent-erect, linear-spathulate lobes 2,5 mm long . . . . . . 45. **C.fortuita**
17 Corolla-tube cylindric or subcylindric, not or only slightly expanded at mouth, sometimes with slightly prominent sinuses:
21 Corolla-lobes more than 5 mm long:
22 Corolla inflation tuberculate within, corolla-lobes narrowly spathulate or linear, but usually with a small apical swelling;

corona with outer campanulate tube, inner lobes arising from within margin of outer tube, ± spathulate . . . .
41(b). **C.decidua** subsp. **pretoriensis**

22 Corolla inflation without tubercles within; corolla-lobes linear, forming a slender beak; corona shortly united at base and confluent with erect-recurved, spathulate-lanceolate inner lobes . . . . . . 46. **C.linearis**

21 Corolla-lobes ±4 mm long, slightly spathulate, pubescent; corona shortly united at base and confluent with erect-recurved, spathulate-lanceolate, elongate inner lobes . . . .
47. **C.tenuis**

10 Corolla puberulous on outer surface; lobes often with slightly pinched-in margins towards base, spathulate above; tube slightly expanded to mouth .
48. **C.pachystelma**

9 Corona campanulate or saucer-shaped at base produced into five outer lobes; outer lobes bilobuled, chute-like or triangular:

23 Outer corona-lobes entire; inner lobes linear to clavate:

24 Corolla-tube puberulous on outer surface; lobes often with pinched-in margins towards base; inner corona-lobes linear-lanceolate above and contracted to a slender base . . . . 48. **C.pachystelma**

24 Corolla-tube glabrous on outer surface; lobes hairy; inner corona lobes linear-spathulate, obtuse . . . . 36a. sp. (Swaziland)

23 Outer corona-lobes deeply bilobed; inner corona-lobes linear-filiform . .
36. **C.paricyma**

6 Basal corolla-inflation oblong-elliptic or bottle-shaped, about as long or sometimes longer than the upper cylindric or funnel-shaped tube; plants mostly pubescent:

25 Corolla-tube glabrous outside or with few hairs; cage ± rounded:

26 Corolla 30-60 mm long, cage more or less circular in outline; leaves ovate-cordate to lanceolate, often coarsely toothed . . . . 49. **C.meyeri**

26 Corolla under 15 mm long, cage often slightly oblique from tube; lobes with long crinkly hairs on basal half . . . . 50. **C.stenoloba**

25 Corolla-tube densely puberulous outside; corolla-lobes and cage much shorter than tube . . . . 51. **C.bonafouxii**

SPECIES DESCRIPTIONS AND ILLUSTRATIONS

1. **Ceropegia bowkeri** *Harv.*, Thes.Cap. 1:9, t.14 (1859); R.A.Dyer in Flora S. Afr. 27,4:49 (1980) [Fig.69, 70]

Perennial herb with cluster of fleshy roots, producing 1-3 annual stems. *Stems* upright, unbranched above, 20-50 mm high. *Leaves* ascending slender, 40-135 mm long, 3 mm broad, without hairs. *Flowers* solitary, lateral at the upper few nodes. *Corolla* 35-40 mm long; tube ascending 16-18 mm long, with inflated base about 6 mm diameter, pale greenish with violet-brown marking at mouth; lobes free pendulous, motile, with row of large pits or depressions

Figure 69. *Ceropegia bowkeri:* a, plant with tuft of fleshy roots and drooping, hinged, corolla-lobes, x ⅔; b, corolla, x 1; c, corolla removed exposing corona, x 4 (after Botanical Magazine plate 5407).

on each side of midrib, light greenish-yellow, shortly hairy on face and with yellow hairs on margin.

Ceropegia bowkeri was described in 1859 in Thesarus Capensis by William Harvey. He named it after Henry Bowker, commander of the Frontier Armed and Mounted Police in Kreilis Country (Transkei). Henry Bowker and his sister Mary Elizabeth (later Mrs F.W.Barber) collected extensively together in the Transkei and they share the credit for many new discoveries. Harvey expressed himself in the following terms about the new *Ceropegia.* 'A very beautiful and curious species for which I am indebted to Mrs Barber, and in giving it her family name, I express my gratitude to the donor as well as the discoverer.'

At the time of the first receipt of the dried specimen Harvey was unaware of the extraordinary character of the corolla-lobes and he illustrated them as erect from the tube. The lobes are in fact, hinged near the base and hang down, swaying to and fro with the slightest breeze. A living plant was illustrated in 1863 correcting the fault in the first illustration when a root flowered at the Royal Botanic Gardens at Kew.

Seed, collected by Mrs Barber in the Transkei on a later occasion, was sent overseas to Dr Moor of Glasnevin, where it flowered in May 1865 and was also painted. It showed sufficient distinction in corolla form from *C. bowkeri* to be given the new name *sororia* (sisterly) by Harvey. The main distinction of *C. sororia* is in the basal inflation of the corolla-tube which passes gradually into the neck whereas that of *C. bowkeri* is globose and the mouth wider. The distinction was accepted until Huber scrapped it in his monograph of the genus in 1958.

In more recent times Larry Leach, under *Leach & Bayliss 12569,* makes the following field note based on several specimens from the veld near Grahamstown which he identified with *C. sororia:* the flowers varied considerably in size and degree of basal inflation but he was of opinion that the two plants should retain specific status as opposed to Huber's treatment of the names as synonyms. The typical form of *C. bowkeri* has not been collected again in over 100 years and there is no record of its degree of variability in the field.

In view of the obvious close relationship shown by the coronal structure of *C. bowkeri* and *C. sororia* it seems that subspecific rank for *C. sororia* meets the case, thus:

Basal inflation of corolla-tube subglobose, passing more or less abruptly into upper cylindric portion — subsp. **bowkeri**

Basal inflation of corolla-tube elongated-ovoid, passing gradually into upper portion and mouth — subsp. **sororia**

subsp. **bowkeri** *R.A.Dyer* in Flora S.Afr. 27,4: (1980) [Fig.69]

Recorded from the Transkei originally well over 100 years ago and no further information added in the interim. Its rediscovery would be a most noteworthy event.

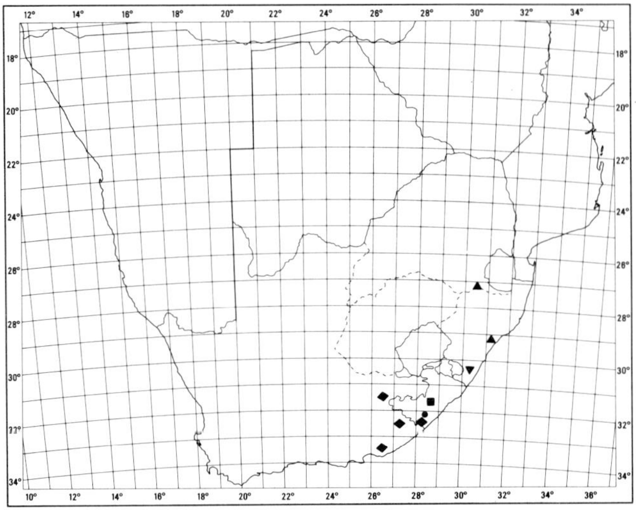

● *Ceropegia bowkeri* Harv. subsp. *bowkeri*
♦ *Ceropegia bowkeri* Harv. subsp. *sororia*
■ *Ceropegia tomentosa* (2)
▲ *Ceropegia scabriflora* (3)
▼ *Ceropegia rudatisii* (4)

Figure 70. *Ceropegia bowkeri* subsp. *sororia:* twig with flower showing bottle-shaped basal inflation and drooping hinged corolla-lobes, x 1 (*Bayliss & Leach* 12569).

subsp. **sororia** *(Harv. ex Hook.f.); R.A.Dyer* in Flora S.Afr. 27,4: (1980) [Fig.70]

Habit as for subsp. *bowkeri. Corolla* 40-60 mm long; tube 25-38 mm long, with elongated ovoid-inflated base 6-8 mm diameter, narrowed into cylindric upper part about 3 mm diameter and expanded to about 5 mm diameter at mouth, variously dotted or violet diffused; lobes with densely greyish white-pubescent band down middle, sometimes giving off transverse green bars on each side, but pattern variable, densely ciliate with purplish hairs.

Plants matching the original one grown from seed collected in the Transkei by Mrs Bowker have since been recorded from the eastern Cape near Queenstown, King Williams Town, Grahamstown, and Bathurst by E.E.Galpin, T.R. Sim, J.Acocks, L.Leach and Roy Bayliss.

2. **Ceropegia tomentosa** *Schltr.* in Bot.Jb.18, Beibl.45:33 (1894); R.A.Dyer in Flora S.Afr. 27,4:49 (1980) [Fig.71]

Perennial herb with a cluster of fleshy roots, producing 1 – few annual stems. *Stems* ±300 mm high, without hairs. *Leaves* slender, ascending up to 100 mm long and 3 mm broad. *Flowers* solitary, lateral at upper nodes. *Corolla* 40-50 mm long; tube 16-18 mm long, with ovoid basal inflation about 5 mm diameter, contracted to 2-3 mm and broadening to mouth about 6 mm wide, greenish yellow with purple-brown markings; lobes free at tips, slender, 25-30 mm long, margins recurved in lower half, densely hairy with short curled or interwoven yellow hairs on base and ciliate with tuft of subclavate red hairs about 5 mm above base.

Ceropegia tomentosa, the hairy one (corolla-lobes), as pointed out by Schlechter when he described this 'interesting and rare' species is related to *C. bowkeri* and the subsp. *sororia.* It is distinguished from them by the linear thread-like (not oblong) corolla-lobes. N.E.Brown adds to this information in the Flora Capensis, citing specimens at the Royal Botanic Gardens, Kew, that Barber & Bowker 372 came from the Transkei near Fort Bowker on the Bashee Ridge and another collection was from near Butterworth. One of these collections is represented in the Albany Museum Herbarium, Grahamstown, a photograph of which is reproduced as the best illustration available.

It must have been relatively uncommon even in the early days of botanical exploration otherwise one would have expected Alice Pegler from Kentani at least to have found it.

Is there sufficient incentive to induce a modern collector to rediscover this species, which apparently has not been collected by other than brother and sister Bowker mentioned under *C.bowkeri.*

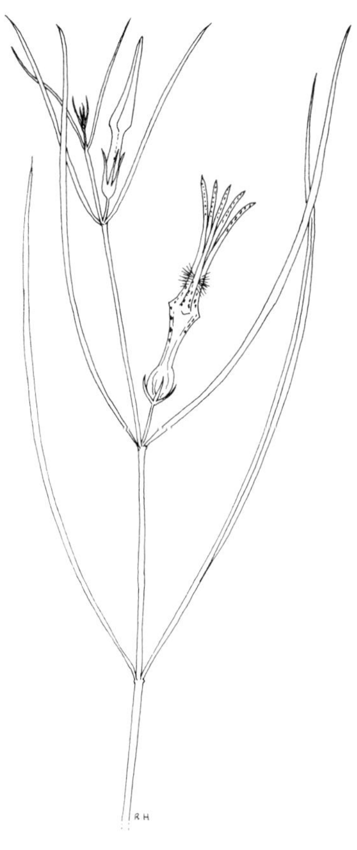

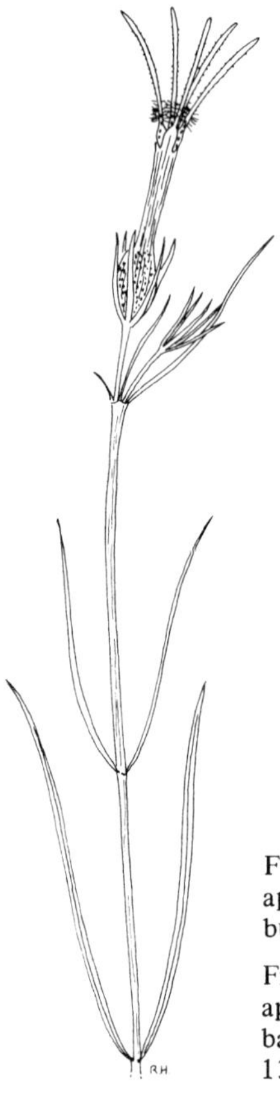

Figure 71 (left). *Ceropegia tomentosa:* apical portion of stem with flower and bud, x ½ (after BRI records).

Figure 72 (right). *Ceropegia scabriflora:* apical portion of stem with inflated base of corolla scabrid (after *J.Acocks* 13158).

3. **Ceropegia scabriflora** *N.E.Br.* in F.C.4,1:810 (1908); R.A.Dyer in Flora S. Afr. 27,4:50 (1980) [Fig.72]

Perennial herb producing a cluster of fleshy roots and a solitary annual stem. *Stems* slender erect, occasionally once-branched above, up to 400 mm tall, without hairs. *Leaves* in 4-7 pairs, slender, ascending 25-50 mm long, with recurved margins. *Flowers* solitary, lateral at upper nodes. *Corolla* 35-65 mm long; tube up to 20 mm long, 6-8 mm diameter at the papillate scabrous, subglobose, inflated base, contracted and smooth above, expanding to about 5 mm wide at mouth; lobes free, slender, up to 35 mm long, with fine hairs on inner face and with tuft of dark purple clavate hairs slightly above base.

Ceropegia scabriflora has a roughly hairy corolla tube and a distinctive tuft of long clavate hairs slightly above the base of the corolla-lobes. This latter feature is shared by the near relative *C.tomentosa* and it is the scabrid basal inflation of the corolla which is the ready means of distinguishing *C.scabriflora.* Time and time again one has to refer to the rarity of preserved records of species of *Ceropegia* (as well as *Brachystelma*) especially among those inhabiting open grassveld. Medley Wood was the sole collector of this species in the previous century near Verulam north of Durban and John Acocks in the present century. Acocks records the following ecological note on his no.13158 in the month of October: 'on quartzite outcrop in subtropical *Hyparrhenia* veld, locally frequent, about 8 km east of Berbice near Piet Retief in southern Transvaal.' Without further positive evidence one may guess but not assume that the species occurred not infrequently between the two proven localities. On the question of colour Acocks noted that the corolla was yellowish-green and purple-striped. The tufts of hair on the lobes were dark purple as described by Brown for Medley Wood's specimen.

Because of their similar habit one feels that *C.scabriflora* is closely related to *C.dinteri* and *C.antennifera*, but there is the fundamental difference in the root system – the former being a cluster of fleshy roots and the latter two with solid tubers.

4. **Ceropegia rudatisii** *Schltr.* in Bot.Jb.40:94 (1907); R.A.Dyer in Flora S.Afr. 27,4:40 (1980) [Fig.73]

Perennial herb producing a cluster of fleshy roots and a single annual stem. *Stem* becoming rather tough, erect, 120-200 mm tall, without hairs. *Leaves* with or without a short stalk, about 60 mm long and 20 mm broad, without hairs. *Flowers* solitary, lateral near the terminal node. *Corolla* 80-90 mm long; tube 25-45 mm long, with inflated base 5-7 mm diameter, reduced to 3 mm and expanded to 7 mm at the mouth, without hairs, but with rows of tubercles within on purple-brown veins; lobes free, slender, slightly spreading, 24-34 mm long with greenish-yellow hairs on inner face and with long clavate vibratile, yellow-based, purple hairs on the margin and face near base.

Ceropegia rudatisii was named after Hans Rudatis a horticulturist-naturalist from Germany. There is little cause to confuse this species, from the Natal coastal region, with any other because of the unique combination of vegetative and floral characters. The flower may be nearly as much as half the length of the full grown stem, while the follicles (fruits or seed pods) are quite as long; the leaves are broad and overlapping. The species has been recorded by no more than four collectors, the last about 50 years ago.

5. **Ceropegia stapeliiformis** *Haw.* in Phil.Mag. 1827:121 (1827), as *stapeliaeformis;* R.A.Dyer in Flora S.Afr. 27,4:50 (1980) [Plate 12, Fig.74]

Perennial succulent with fibrous rootsystem. *Stems* fleshy, up to about 15 mm thick, mainly decumbent or trailing, rooting in contact with soil, cylindric,

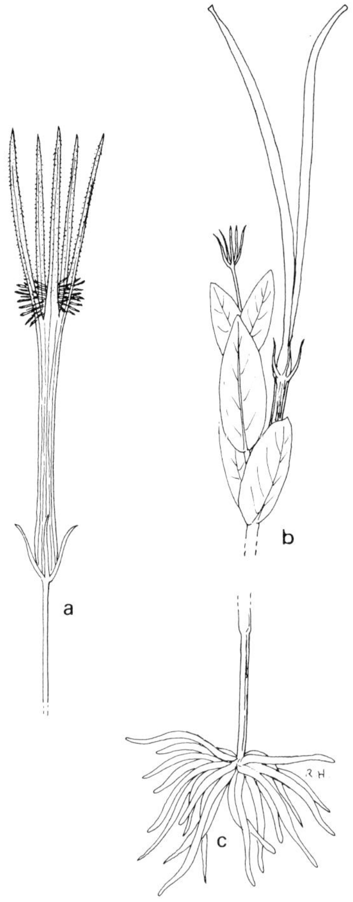

Figure 73. *Ceropegia rudatisii:* a, flower with remarkably long corolla with long clavate, yellow-based motile hairs near base of lobes, x ½; b, fruiting branch with stalk (pedicel) very much thickened after fertilization, x ½; c, rootstock with tuft of succulent or slightly fleshy roots (adapted from *Thode* 259a).

tuberculate; branches constricted at junction, tapering and sometimes climbing towards the inflorescences, glabrous dull green mottled and often tinged with purple. *Leaves* rudimentary, arising from tubercled base, up to 3 mm long, with minute globular swellings at their base, soon falling. *Flowers* in succession from short flowering branches on ultimate growth.

Ceropegia stapeliiformis, so named because of its vegetative similarity to some species of the *Stapelieae,* has been a favourite subject for cultivation from the time of its introduction to Europe over 150 years ago. The spelling of the specific name *stapeliiformis* is in accordance with recommendation 73G. of the International Botanical Code of 1972.

The specimen first described by Haworth in 1827 arrived at Kew, England, in a consignment of succulent plants from their paid collector Bowie at the Cape. It flowered at Kew in July 1826. The colour and sprawling habit of the plant prompted Haworth to refer to it as the 'lurid trailer'. According to N.E. Brown in Flora Capensis 1908, specimens had been preserved many years ear-

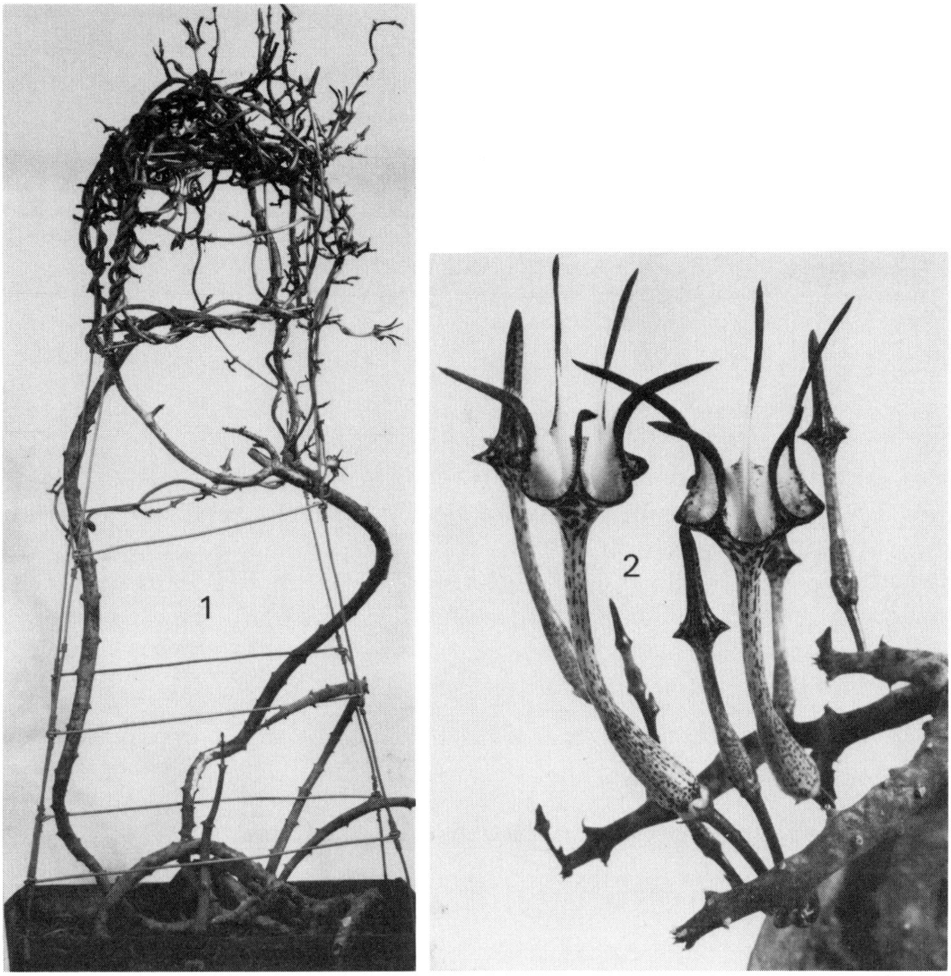

Figure 74. *Ceropegia stapeliiformis* subsp. *stapeliiformis:* 1, plant in cultivation at BRI from cutting collected in karoid scrub near Port Elizabeth, x 1/10 (*D.Comins*); 2, plant from Beaufort West (furthest west locality for species, x 1 (*D.Hardy & R.Bayliss* 996).

lier by Carl Thunberg during his botanical travels in the Cape, 1772–1775.

In the wild the plants are usually associated with scrub bush, and over the years it has been recorded from near Beaufort West eastwards to Humansdorp, Uitenhage, Cradock, Albany and the area of King William's Town. There is no record of any related form from the Transkei or Natal but a close relative appears in Zululand, Swaziland and northern Transvaal. This isolation supported by some differences in character led to the northern plant being given specific rank under the name *Ceropegia serpentina* E.A.Bruce. The author described how the vegetative branches form a straggling snake-like mass on the ground, whilst the longer more slender twining branches climb among bushes and produce flowers on the young growth which breaks through the shade into the sunlight.

The similarities of the two forms were so obvious and the differences somewhat unconvincing that it has been decided to observe only subspecific status on the following lines:

Corolla-lobes spreading-recurved; flowering branches usually not much twining; leaves rudimentary with minute globular stipules at base; outer corona about 2 mm high with bifid lobes ... subsp. **stapeliiformis**

Corolla-lobes not greatly spreading, often cohering and twisted towards apex; flowering branches usually twining; leaves rudimentary, with two golden-yellow glands at base and two on stem near leaf base; outer corona cup-shaped, about 2,5 mm high, lobes bifid ... subsp. **serpentina**

subsp. **stapeliiformis** *R.A.Dyer* in Flora S.Afr. 27,4:51 (1980) [Fig.74]
Corolla tip to about 70 mm long, nearly half-divided; tube ovoid-inflated at base, 6-7 mm diameter, narrowed above, straight or curved, expanded into funnel-shaped mouth, pubescent inside at the mouth and basal part with white, curly hairs, pale greenish white, spotted with violet; lobes free, usually spreading-recurved, with margins folded backwards, 20-35 mm long, pubescent on the lower part of the inner face, with the margins and apical part violet-purple, greenish yellow or purple-brown.

Distribution records are incorporated above.

subsp. **serpentina** *(E.A.Bruce) R.A.Dyer* in Flora S.Afr. 27,4:51 (1980)
The circumstances and characters leading to the recognition of this as a sub-

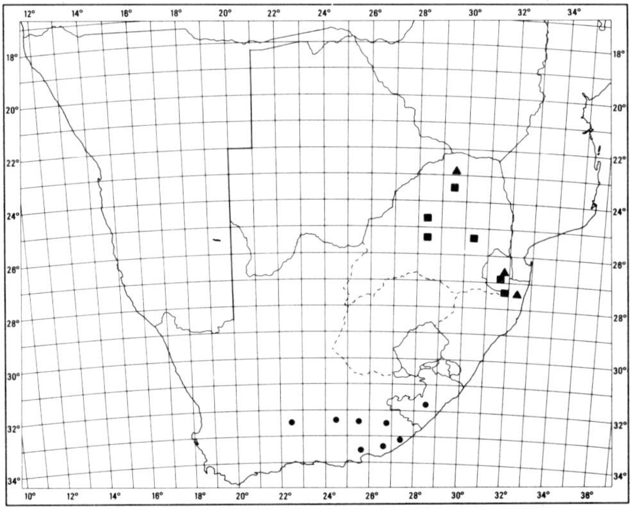

● *Ceropegia stapeliiformis* subsp. *stapeliiformis*
■ *Ceropegia stapelliformis* subsp. *serpentina*
▲ *Ceropegia cimiciodora* (6)

Plate 12. *Ceropegia stapeliiformis* subsp. *stapeliiformis* (species 5); a. flowering branch with secondary rooting, x 1; b. leaf with glands at base, x 1; c. flowering stem, x 1; d. longitudinal section of corolla, x 1½; e. corona forming deep pockets within and incumbent spreading inner lobes, x 10; f. pollinia and shortly winged carrier, x ±40; g. seed with parachute of hairs (*Auriol Batten,* King William's Town)

a
b
c
d
e
f
g

e
c
d
a
b
Auriol Batten

species have been set out above and in the Flora of S.Afr. referred to above. One can only reiterate that the successful cultivation of the subspecies gives great pleasure to the grower.

Figure 75. *Ceropegia cimiciodora:* a, flower and bud side view, x 2; b, flower and bud face view, x 2⅔ (*Obermeyer, Schweickerdt & Verdoorn* 322, from type plant collection north of Zoutpansberg, northern Transvaal, photo H.Lang).

Plate 13. *Ceropegia ampliata* (species 9); a. rootstock with fleshy roots, x 1; b. flowering branches, x 1; c. longitudinal section of corolla, x 1½; d. corona, x 10; e. pollinia with long pellucid inner margin, shortly attached to winged carrier, x 50 (*Auriol Batten,* 321, Fish River Valley, near Peddie, eastern Cape)

6. **Ceropegia cimiciodora** *Oberm.* in Flower.Pl.S.Afr. 13:t.488 (1933); R.A. Dyer in Flora S.Afr. 27,4:51 (1980) [Fig.75]

Perennial succulent with fibrous roots. *Stems* fleshy, tuberculate, shiny, greyish, mottled with olive green, without hairs, scrambling but given the opportunity climbing and twining to several metres in length, sparsely branched. *Leaves* rudimentary, triangular or sub-heart-shaped. *Flowers* 1-2 together opening in succession. *Corolla* about 50 mm long, curved, without hairs, grey mottled and speckled with purple at the apex; tube with small white ring at the base within inflated base, narrowed above to about middle and then expanded to the funnel-shaped mouth, hairy within base and at mouth; lobes free, with the top star-shaped from above, and 20 mm diameter, individually like the inside of a donkey's ear.

Ceropegia cimiciodora was reported first by a trio of botanists, Obermeyer, Schweickerdt and Verdoorn, who made a small collecting expedition to farms north of the Zoutpansberg during the last fortnight in November 1932. One of their prizes was this, then unknown, species of *Ceropegia.* They located it on the farm Chopudi between Waterpoort and Zoutpan creeping and scrambling under the protection of thorn scrub; but it was their entomological colleague, G.van Son, who grew the specimen for description and illustration. The specific name is hardly complimentary to the remarkable flowers which happen to emit the odour of a bedbug (*Cimex*) to attract the fertilizing insect. In the intervening years the species has been traced from the northern Transvaal to Swaziland and Zululand.

Ceropegia cimiciodora is allied to *C.stapeliiformis* in habit but differs markedly in the shape of the flowers. The corolla has a long tube and short, broad lobes, the tips of which become free but do not spread. This is a case where a good illustration is worth many words of explanation.

7. **Ceropegia barbata** *R.A.Dyer* in Bothalia 12,4:000 (1978); Flora S.Afr. 27,4: 53 (1980) [Fig.76]

Perennial half-succulent with a tuft of fleshy roots. *Stem* (annual?) up to about 450 mm long, not visibly twining, somewhat fleshy, sparingly branched, without hairs. *Leaves* slender, 30-60 mm long, somewhat fleshy, with margins probably somewhat recurved. *Flowers* solitary, few, lateral at nodes. *Corolla* about 25 mm long; tube about 13 mm long, straight or slightly curved with globose-elliptic inflation at base, 5-6 mm diameter, narrowed to about 3 mm and then expanded to the mouth 8-9 mm diameter, longitudinally (purple?) veined; lobes twisted in bud up to 12 mm long, with terminal knob, becoming free, spreading erect, slender, with long cylindric dark-tipped hairs from the

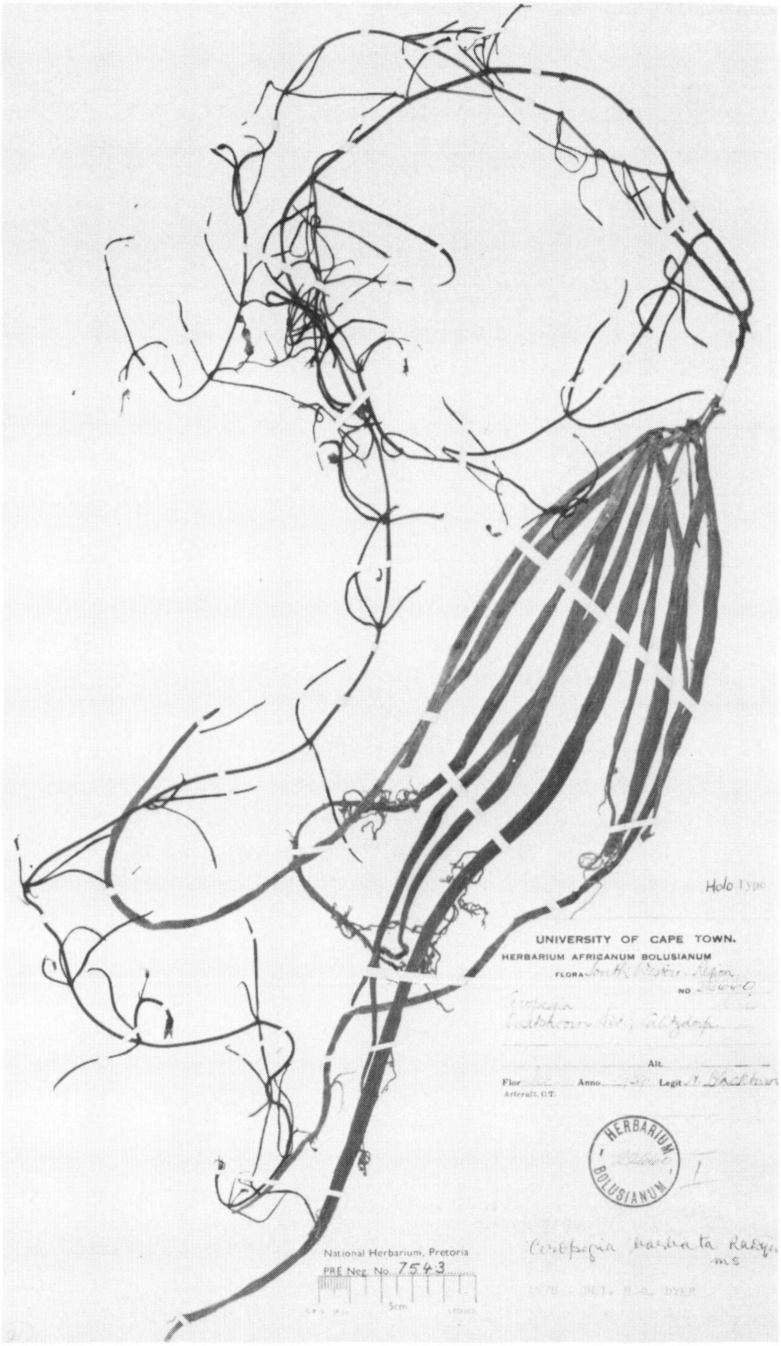

Figure 76. *Ceropegia barbata:* photograph of type specimen in Bolus Herbarium from near Calitzdorp in the Oudtshoorn district, Cape, x ⅓ (*Blackburn* Bol. 22660/1).

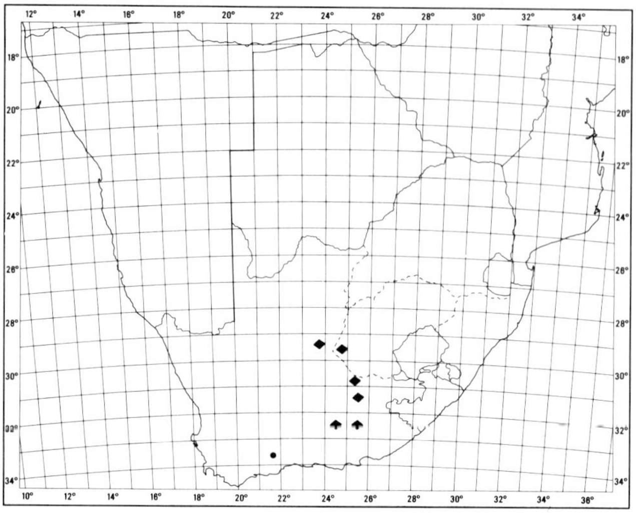

● *Ceropegia barbata* (7)
♦ *Ceropegia filiformis* (8)

inner surface and with a mop of long filiform hairs from the somewhat club-shaped apex.

Ceropegia barbata with bearded corolla-lobes is based on specimens which remained in a botanical institution unidentified for 40 years after collection. They were sent to the Bolus Herbarium, University of Cape Town, by S.Blackburn from near Calitzdorp in 1939. They constitute the only record even now. H.Blackburn was either magistrate or station master of the small Cape village whose surroundings are noted for many unusual succulent plants mixed with karroid scrub. There are probably more unrecorded treasures in that environment.

Ceropegia barbata is distantly related to *C.filiformis* but differs in the corolla-lobes being free at the tips on opening and they are furnished with distinctive long pubescence, the specific name being in reference to these hairs.

8. **Ceropegia filiformis** *(Burch.) Schltr.* in Bot.Jb.2 Beibl.54:13 (1896); R.A. Dyer in Flora S.Afr. 27,4:53 (1980) [Fig.77]

Perennial herb with rootstock producing a cluster of fleshy roots. *Stem* annual up to about 800 mm long, twining, sparingly branched, without hairs. *Leaves* slender, up to 70 mm long and 10 mm broad, herbaceous. *Flowers* lateral from upper nodes, 1-3 developed successively. *Corolla* up to 60 mm long; tube up to 40 mm long, straight or slightly curved, with oblong-elliptic basal inflation up to 12 mm diameter, narrowed to about 3 mm and then expanded to 8-9 mm at the mouth, tuberculate on veins within inflation, pale-green, longitudinally veined with dull purple; lobes twisted in bud, up to 20 mm long, spreading

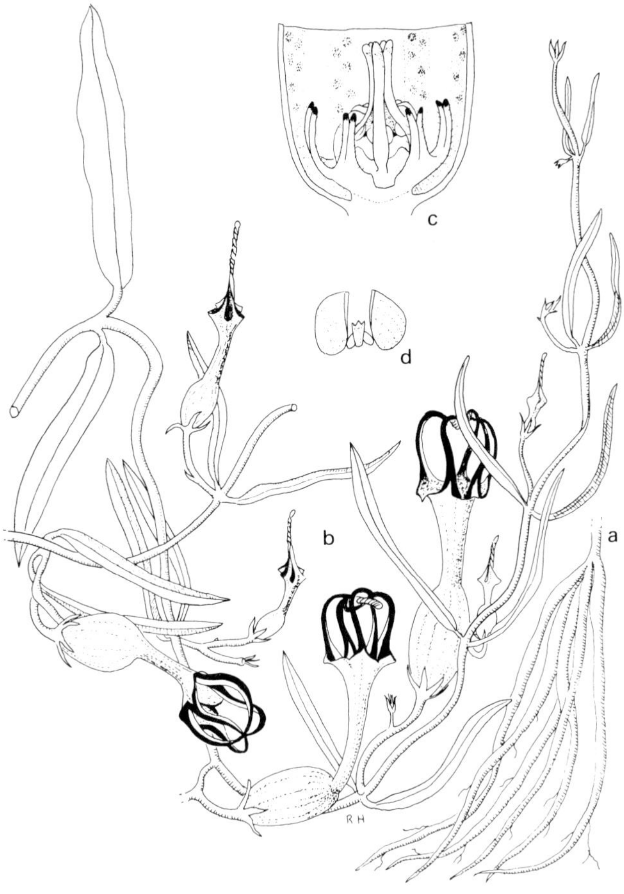

Figure 77. *Ceropegia filiformis:* a, tuft of fleshy roots, much reduced; b, flowering stem, x ½; c, section of base of corolla exposing corona, x 5; d, pollinia with short caudicles (attachments) to shortly winged carrier; x 30 (after *T.G.la G.Joubert* in PRE 30356).

erect except for connate twisted tips 2 mm long, white keeled down inner face of lower half, otherwise strikingly dark purple.

Ceropegia filiformis, in the shape of its expanded corolla-lobes, reminded the famous traveller, William Burchell, of a royal crown when first he saw a specimen one early morning in 1812 near his camp in Griqualand West. The tube is funnel-shaped. Records in the field to date have established the distribution range to extend in Karooveld from near Aberdeen and Cradock in the eastern Cape to the more sandy northern parts. It is usually associated with low shrubs which afford it protection and support.

9. **Ceropegia ampliata** *E.Mey.* in Comm.Pl.Afr.Austr.194 (1837); R.A.Dyer in Flora S.Afr. 27,4:54 (1980) [Plate 13, Fig.78]

Perennial succulent with a cluster of fleshy roots or with rather fibrous roots from rooted stems. *Stems* succulent, rambling, twining up to about 2 m and

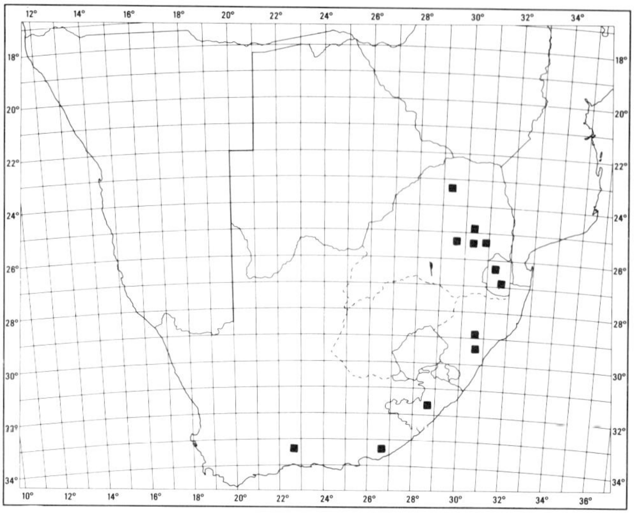

■ *Ceropegia ampliata* (9)

more in length, leafless and without hairs. *Leaves* rudimentary on terminal growth, soon falling. *Flowers* 2-4 together, lateral at the nodes, developed in succession. *Corolla* 50-70 mm long; tube straight or slightly curved, 35-50 mm long, with subglobose basal inflation up to about 20 mm diameter, 22 mm in length, narrowed to 12-13 mm diameter, and cylindric up to the mouth, pale green with narrow purple band round mouth of inflation, with long hairs mainly round mouth and above inflation; lobes 10-20 mm long, tapering from broad base, with replicate margins curved to the normally united tips, mainly bright green and veined.

Ceropegia ampliata is so named because of the generous proportions of the corolla tube which is among the broadest in the genus. The German collector J.F.Drège gathered the first specimens in the scrub bush of the lower Great Fish River valley about 1830. At the same time he collected *Ceropegia carnosa* and the far rarer *C.fimbriata. C.ampliata* is now known to occur more or less continuously from the southern Cape near Oudtshoorn and Graaff Reinet into Natal and northwards into Zululand and Transvaal, then it takes another jump into tropical Africa and Madagascar. In tropical Africa it bears the varietal name *oxylola* H.Huber and in Madagascar the subspecific name *madagascariensis* Lavranos.

In observing specimens collected by Auriol Batten from the type locality in the lower Great Fish River Valley and also from near Graaff Reinet it seems that the connective of the tips of the corolla lobes may be very delicate. The connective may rupture and the lobes become free on the living plant and the lobes of all flowers may separate when placed in preservation or on pressing for herbarium records.

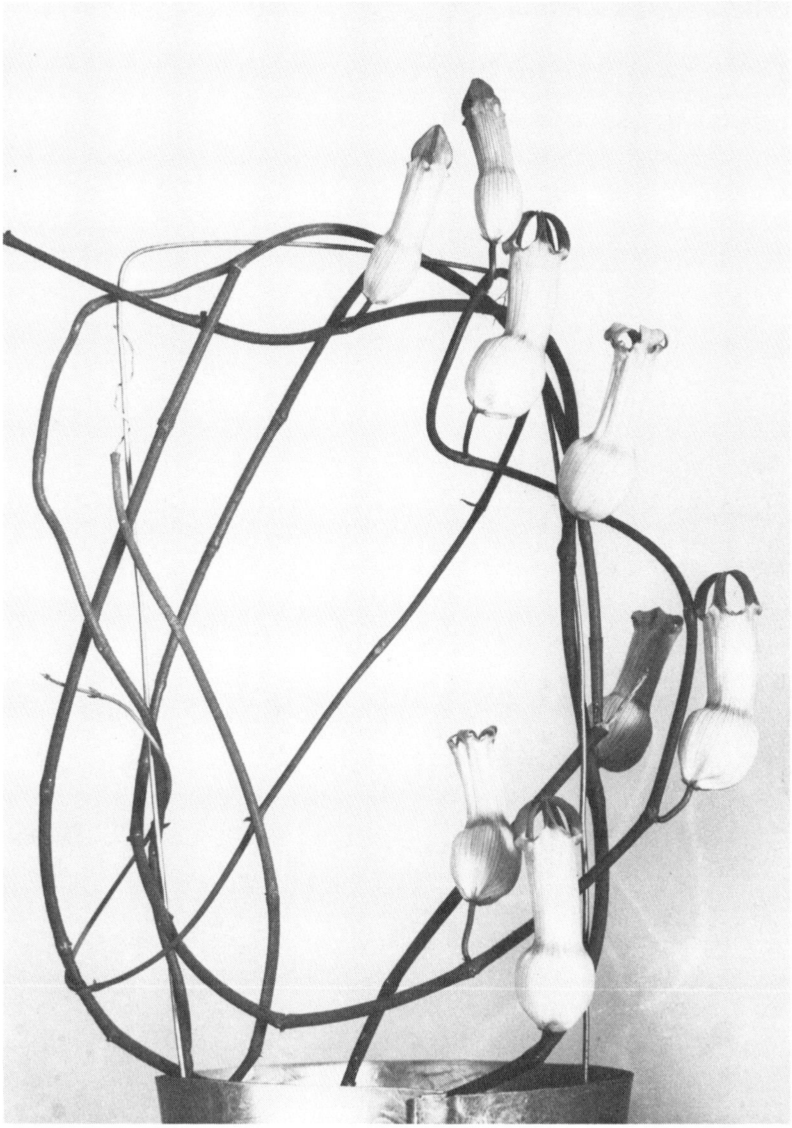

Figure 78. *Ceropegia ampliata:* typical floriferous habit with succession of open flowers, x ½ (*R.Bayliss* 313, near Gollel, Swaziland, cult. BRI, photo J.Reyburn).

It is revealed by various illustrations and dissections that one has taken the uniformity of *C.ampliata* too much for granted. In actual fact there is considerable variation in robustness, size of flowers and even in the shape of the corona. The corona may be basin-shaped or shallower and more solid from which foundation the inner filiform lobes arise to about 5 mm high.

10. **Ceropegia fimbriata** *E.Mey.*, Comm.Pl.Afr.Austr. 194 (1837); R.A.Dyer in Flora S.Afr. 27,4:54 (1980) [Fig. 79]

Perennial with slender leafless stems and a tuft of slender succulent roots. *Stems* scrambling, climbing and twining up to about 1 m or more long, moderately branched. *Leaves* rudimentary, slender not more than 5 mm long. *Flowers* 1-2, lateral at the nodes. *Corolla* up to about 60 mm long; tube about 37 mm long, slightly curved, with a subglobose inflated base 12-13 mm long, 10 mm diameter, contracted to 3 mm and thence expanded to mouth about 15 mm diameter, without hairs mainly green; lobes strut-like, slender upright, in basal 4 mm, expanded and united into an umbrella or parachute-like canopy 16-18 mm diameter, with 10-crenate margin, fringed with long, pendulous, clavate, vibratile, purple hairs.

Ceropegia fimbriata so named, because of the fringe of clavate-vibratile hairs dangling from the rim of the corolla-canopy, must have given Drège, who also collected the previous species *C.ampliata,* much pleasure when he discovered it in scrub bush on the banks of the Great Fish River Valley near Trumpeter's Drift in the eastern Cape. In spite of fairly intensive collecting in tnis environment, *C.fimbriata* has been found only rarely, while *C.ampliata* occurs

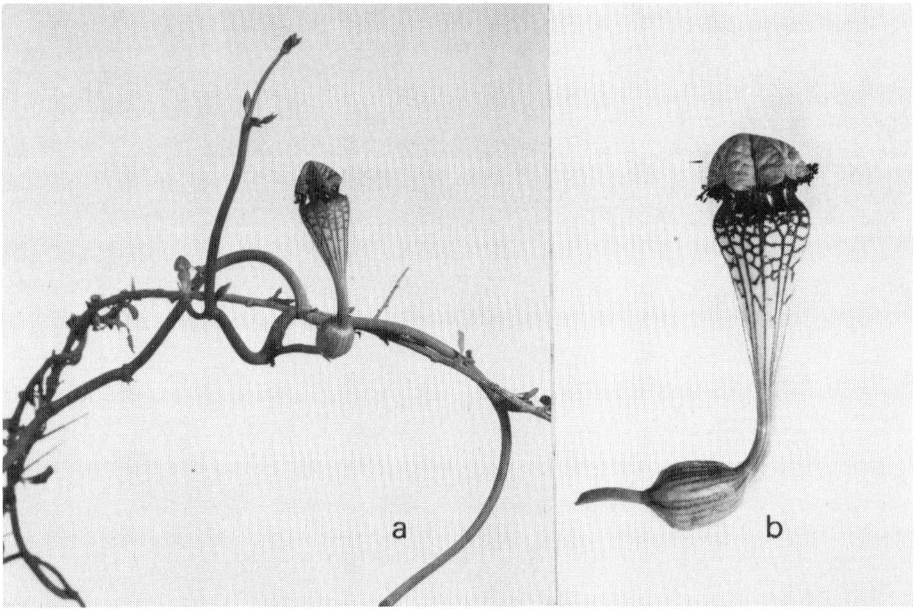

Figure 79. *Ceropegia fimbriata:* a, habit, reduced; b, flower with long, clavate, vibratile hairs pendulous from the canopy, x 1 (*Estelle Bayliss,* 1979, Fish River Valley scrub some miles northeast of Grahamstown, Cape).

far and wide. When Estelle Bayliss rediscovered *C.fimbriata* north of Grahamstown in 1964 over 100 years after Drège, it was mistakenly identified as an undescribed species.

11. **Ceropegia connivens** *R.A.Dyer* in Bothalia 12,4:630 (1979); in Flora S. Afr. 27,4:54 (1980) [Fig.80]

Perennial with tuft of fusiform roots and sometimes stems rooting in contact with soil. *Stems* slender, succulent, sparingly branched, weakly climbing, perennial without hairs. *Leaves* rudimentary. *Corolla* 30-40 mm long, without hairs on outer surface; tube with basal globose inflation about 7 mm diameter, contracted above and then gradually expanded to mouth, conspicuously veined;

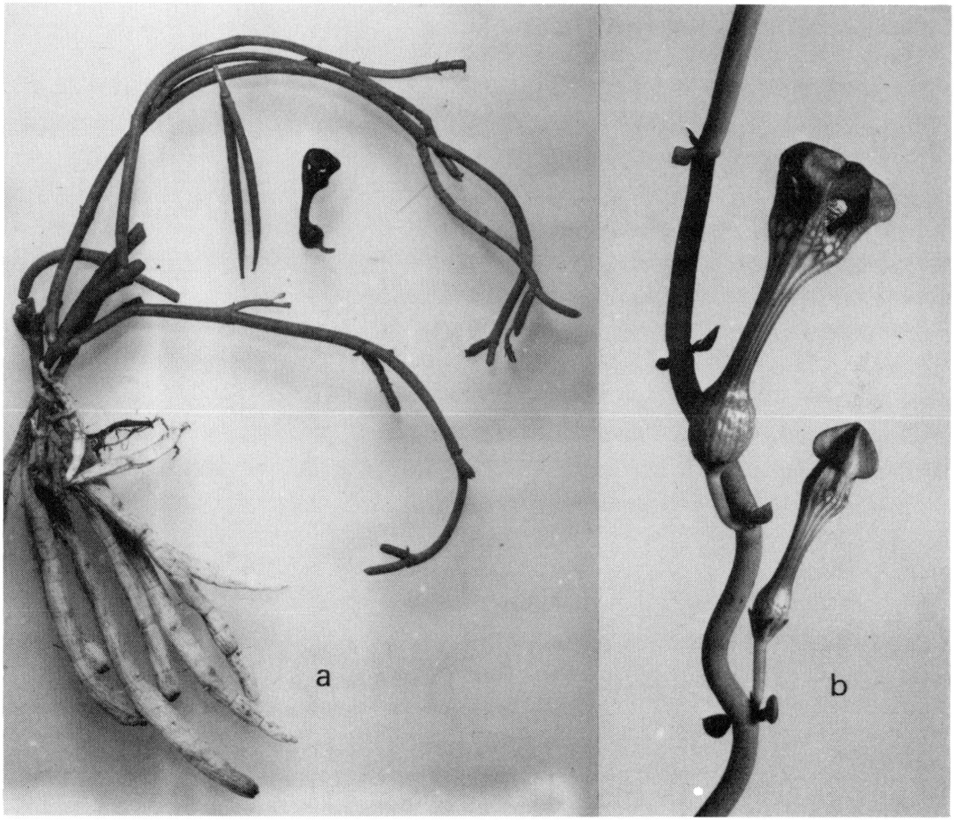

Figure 80. *Ceropegia connivens:* a, type habit, with fleshy roots, young seed pods and separate flower, x ½; b, flowering stem with open flower but long clavate hairs directed inwards and not clearly visible, x 2 (*F.Stayner* KG59/490; Fig.3 in Bothalia 12,4:630 (1979)).

lobes united into an incised canopy 10-12 mm diameter, more or less flat-topped, broader than the mouth of the tube, with long clavate hairs mostly directed inwards from the margins.

Ceropegia connivens, so named because the long hairs on the margins of the corolla-lobes are normally directed inwards towards each other within the cage-like structure. In pressed specimens, however, some of the dense cilia of hairs may be forced outside the corolla openings. In 1975 P.Bruyns collected seed in the wild near Worcester which was germinated at the Worcester Karoo Garden. One seedling donated to the Botanical Research Institute, Pretoria, flowered continuously throughout January 1980 when it was illustrated for publication. It differed, however, in a smaller canopy to the corolla and some of the long, somewhat vibratile hairs became exserted from between the canopy struts. It is regarded as a 'form' of the parent species.

One would expect that the newly described *C.connivens* and *C.geniculata,* with headquarters near Worcester and Ladismith would have been well recorded in the early days of exploration because of their proximity to the Cape. It seems that the early botanists at the Cape were so overwhelmed by the richness of the Flora that they collected the more conspicuous components nearest to hand and had little incentive to specialize at that stage of exploration. In present circumstances specialization comes naturally and plants overlooked by the early explorers, became obvious to the specialist.

The first known record of this species in the Marloth herbarium was mistakenly identified as *C.fimbriata* which caused misunderstandings for some years. It may be of interest to observe that *C.occulta,* no. 40, with a similar growth habit, but with a tuberous rootstock and no long hairs on the corolla-lobes, also occurs near Worcester.

12. **Ceropegia geniculata** *R.A.Dyer* in Bothalia 12,4:631 (1979); in Flora S. Afr. 27,4:55 (1980) [Plate 14, 14a, Fig.81]

Perennial with a tuft of fusiform roots, sometimes re-rooting. *Stems* slender, succulent, climbing, sparingly-branched, without hairs. *Leaves* rudimentary. *Corolla* 40-60 mm long; tube globose-inflated at base, gradually expanded to mouth, without hairs; lobes 15-25 mm (30) long; spathulate towards apex, 3-7 mm broad, variably connate towards apex, with margins more or less recurved, kneed and purple-coloured from near the middle to near the base, usually with an abundance of long clavate hairs spreading in all directions.

Ceropegia geniculata, although only recently distinguished and described, has been recorded from several fairly widely separated localities in a relatively short space of time. It is distributed in scrub bush in the Little Karoo from Ladismith to Oudtshoorn in the Cape. One question is why wasn't it collected

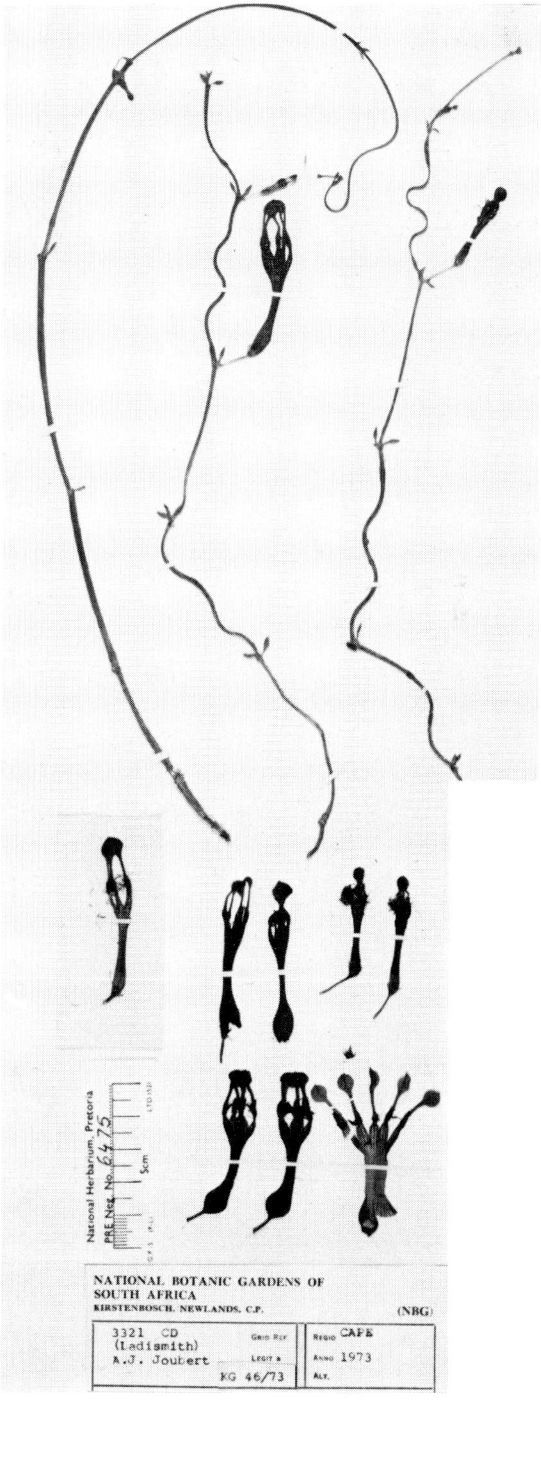

Figure 81. *Ceropegia geniculata:* holotype, corolla-lobes 'kneed' at area with long clavate hairs, x 0,5 (Fig.4 in Bothalia 12,4:631 (1979)).

earlier even though it is not readily detected when mingled with low Karoo bushes. Another question is the reason for the relatively wide range of variation in the corolla. The lobes or struts are short and broad or long and narrow: long purple vibratile hairs are present on the short to medium-length struts but not from the most slender long lobes: the knee in the lobes varies in position from near the middle to near the base and its regular presence is the unifying character of the species concept and the reason for the specific name. One may add fuel to the fire of speculation by mentioning that the related species *C.zeyheri* has been recorded from within its distribution range and that another relation *C.connivens* occurs a little further west in the Worcester district, but so far they have not been recorded together in the wild.

13. **Ceropegia zeyheri** *Schltr.* in Bot.Jb.38:48 (1905); R.A.Dyer in Flora S.Afr. 27,4:55 (1980) [Plate 14a, Fig.82]

Perennial with a cluster of soft fleshy roots and a slender leafless stem. *Stem* succulent, sparingly branched, scrambling or climbing-twining, without hairs. *Leaves* rudimentary. *Flowers* 1-3, lateral at the nodes, developing in succession. *Corolla* 30 to nearly 60 mm long; tube more or less globosely inflated at base to 5-8 mm diameter, subcylindric above and slightly expanded to the mouth, without hairs externally; lobes linear, connate at the apex forming a slender cage-like structure, sometimes with few hairs outside for short distance near base, then sometimes with patch of long, clavate-vibratile, caducous hairs and shortly hairy above, margins infolded.

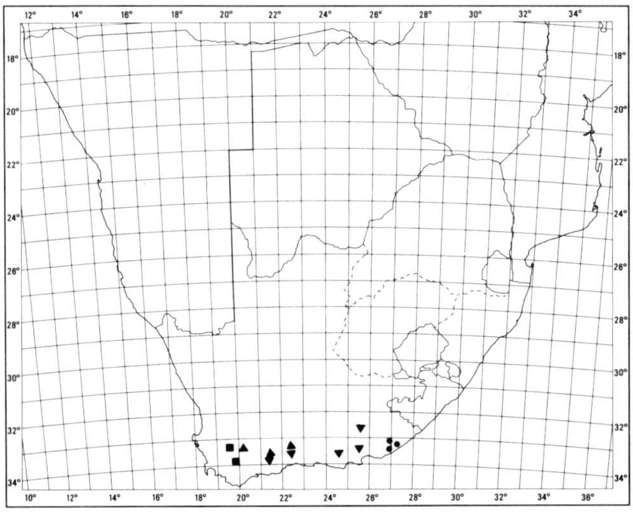

● *Ceropegia fimbriata* (10)
■ *Ceropegia connivens* (11)
▲ *Ceropegia geniculata* (12)
▼ *Ceropegia zeyheri* (13)

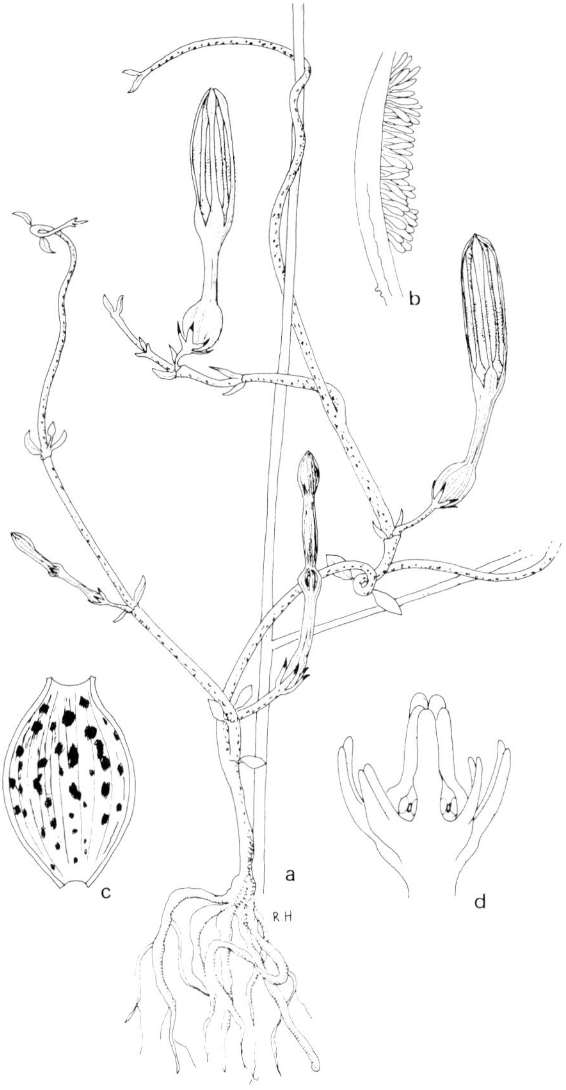

Figure 82. *Ceropegia zeyheri:* a, typical form in scrub bush on old road from Uitenhage to Port Elizabeth, x ½; b, long clavate, vibratile hairs on inner face of corolla-lobes, x 5; c, internal view of inflated base of corolla tube, x 2; d, corona x 5 (after Flower. Pl. Afr. t.1296).

Ceropegia zeyheri, named after one of the lynx-eyed trio of German collectors of the early 1800's, Ecklon, Zeyher and Drège. When first describing the species Schlechter mentioned that it was found in Karoo scrub near the Great Vaal River, Transvaal, but, as pointed out by N.E.Brown in Flora Capensis (1908), the Great River or Groot Rivier in Zeyher's time referred to the Gamtoos River of the eastern Cape. Unfortunately this type specimen was destroyed in the Berlin herbarium and Huber selected another of Zeyher's specimens from the Karoo scrub bush of the Swartkops River near Uitenhage to represent the species. The species has since been found in a much wider area of distribution as far west as near Ladismith in the Cape.

It is normal to find a fair divergence in habit of a widely spread species and one must allow for this in *C.zeyheri.* It may or may not have climbing tendencies, but what is less understandable is the presence or absence of a tuft of clavate-vibratile hairs from the inner surface of the corolla-lobes a little distance from the base: when present they seem to fall away rather readily. The length of the corolla and of the lobes is also very variable.

In recent times allied species have been distributed, including *C.geniculata,* which have overlapping distribution records and one has in mind the possibility of natural hybridization when working in the field. Plants may thrive in cultivation for a season or two but it requires special intuition to maintain them for any length of time.

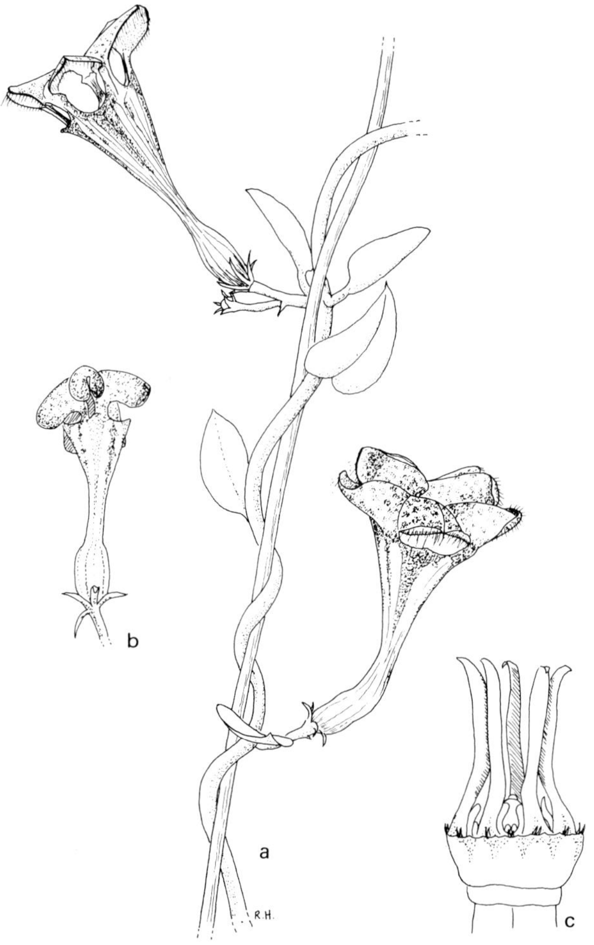

Figure 83. *Ceropegia sandersonii:* a, habit of more or less typical form, x ½; b, bud with strange appearance of canopy in bud, x ½; c, corona with outer basin-shaped base and erect linear-filiform inner lobes, x 5 (after *L.E.Codd* 7127, 7642).

14. **Ceropegia sandersonii** *Decne. ex Hook.f.* in Curtis's bot.Mag.t.5792 (1869); R.A.Dyer in Flora S.Afr. 27,4:56 (1980) [Fig.83, 83A]

Perennial, somewhat succulent, producing a cluster of long cylindric fleshy roots. *Stem* perennial, twining up to about 2 m tall or longer, about 5 mm thick, fleshy, without hairs. *Leaves* 20-50 mm long, 12-25 mm broad. *Flowers* 2-4 in stalked cymes, developed in succession. *Corolla* 40-70 mm long; tube 25-50 mm long, with elongated basal inflation ±6 mm diameter, narrowed and usually somewhat curved above, and expanded to the funnel-shaped mouth 15-25 mm diameter, with few hairs near base within, light green on inflation, striped with green and white above; lobes united above and opening into a 5-keeled concave, flattish, or umbrella-like canopy 25-50 mm diameter, apiculate from the centre, supported by five broad struts with wide openings be-

Figure 83A. *Ceropegia sandersonii:* form once distinguished under the name *Ceropegia monteiroae,* x ¾ (*C.J. Ward* 3825, Ndumu Game Reserve, Zululand).

tween; margin of canopy with long white to purple, sometimes vibratile hairs.

Ceropegia sandersonii is named after John Sanderson, journalist in Natal, with many talents and a keen interest in Botany. He first collected plants in 1867 on the banks of a stream leading into the Umgeni River and forwarded a specimen to Kew in the following year. When describing the form *Ceropegia monteiroae* in 1887 J.D.Hooker mentioned Sanderson's plant, growing in a pot in the Succulent House at Kew, and trained up a rafter where it attracted attention by its curious green flowers expanding into the form of a trumpet surrounded by a canopy fringed with flat erect hairs and supported on five short legs.

There can be little argument that this is one of the more striking species among many unusual growth forms in the genus. It has an oriental look about the corolla canopy with either erect or pendulous hairs from the margin. In the form, at one time named *C.monteiroae,* the canopy is dome-shaped and the cilia pendulous and purple but all intermediate gradations occur.

Roughly the distribution extends from the warm Natal coastal belt northwards into Zululand, Maputo, Swaziland and hot eastern Transvaal.

15. **Ceropegia distincta** *N.E.Br.* in Kew Bull. 1895:262 (1895); R.A.Dyer in Flora S.Afr. 27,4:56 (1980) [Fig.84, 84A,B]

Perennial with rootstock producing a cluster of fibrous to fleshy roots. *Stem* subsucculent, twining, up to 2 m or more long, without hairs. *Leaves* somewhat fleshy, drying thin, up to 75 mm long, 50 mm broad. *Flowers* 1-2 (few) at the node, pedunculate. *Corolla* up to 50 mm long; tube 25-30 mm long,

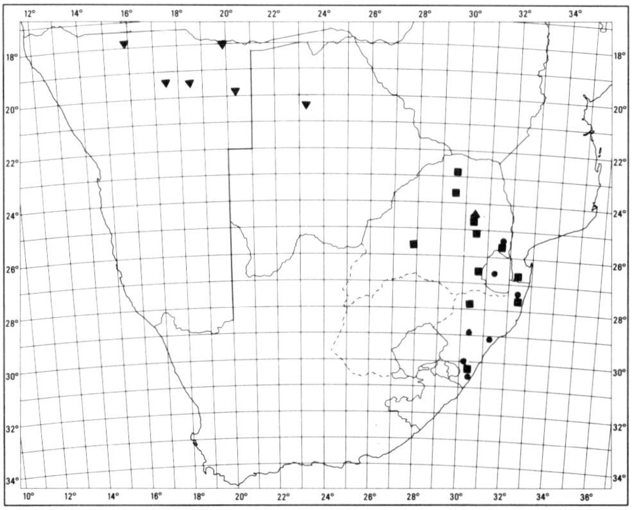

● *Ceropegia sandersonii* (14)
■ *Ceropegia distincta* subsp. *hygarthii*
▲ *Ceropegia distincta* subsp. *verruculosa*
▼ *Ceropegia lugardae* (16)

Plate 14a. Three flowers on the right: *C.zeyheri*, two flowers on the left: *C.geniculata*

Plate 14. *Ceropegia geniculata* (species 12); specimen from Van Wyk's Dam, Ladismith, Cape Province, A.J.Joubert, cultivated at Karoo Garden, Worcester 46/73, photo: B.Bayer)

a
b
c
d
e
f
g
Auriol Batten

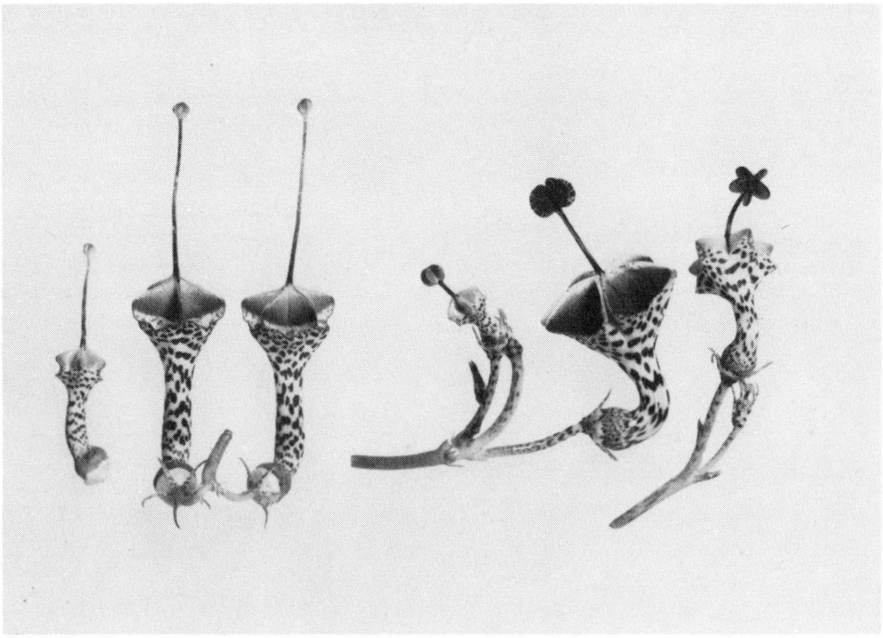

Figure 84. *Ceropegia distincta:* range of variation in the corolla form and apical knob, x 1 (after BRI records).

with slightly globose base and curved upwards nearly at right angles, cylindric and expanding to 15-20 mm at mouth, without hairs outside, hairy at base and in throat within; lobes more or less triangular at base united above into a short or long slender tube 5-15 mm long and abruptly dilated at apex into a 5-winged knob (1)2-5 mm diameter.

Ceropegia distincta even in its several forms, is still distinctive. The type material from Zanzibar consists of an immature flower damaged by insect attack. It has unusually long calyx lobes and large leaves, which seem in keeping with such a tropical environment. In Southern Africa, before the name change was introduced by Huber, it was known under the name *C.haygarthii* and it has been found desirable to adopt subspecific status for two southern forms on the following comparisons:

Corolla-lobes with a deltoid-ovate base, connivent above and connate into a slender tube 5-15 mm long and abruptly dilated at apex into a 5-winged knob, 3-5 mm diameter; stems smooth; roots fibrous subsp. **haygarthii**

Plate 15. *Ceropegia crassifolia* (species 23); a. rootstock with cluster of fleshy roots and basal leaf of a annual growth, x 1; b. flowering stem, x 1; c. fruit pods or follicles which develop in pairs from single flower, x 1; d. colour variation of flower, x 1; e. longitudinal section of corolla showing internal hairs in tube and on lobes, x 1½; f. corona with 5 broad outer lobes confluent with base of inner incumbent lobes, x 10; g. pollinia attached by slender stalks (caudicles) to winged carrier, x ±40 (*Auriol Batten,* King William's Town Commonage)

Figure 84A. *Ceropegia distincta* subsp. *verruculosa:* typical form showing warty stems and corollas with cage-like tips, from near Burgersfort, NE Transvaal, x ¾ (*D.Roux* 2).

Corolla-lobes with a deltoid-ovate base, connate above into a slender tube and becoming free and forming a narrow terminal cage; stems verruculose warty or roughly pimpled roots fleshy subsp. **verruculosa**

subsp. **haygarthii** *(Schltr.) Huber* in Mems Soc.broteriana 12:90 (1958); R.A. Dyer in Flora S.Afr. 27,4:57 (1980) [Fig.84B]
Ceropegia distincta subsp. *haygarthii* was named after W.J.Haygarth who had given his uncle J.Medley Wood a root for cultivation at the Natal Botanic Garden, Durban. The distribution of this subspecies is not unlike that of *C.sandersonii,* sp.no.14, in that it occurs mainly in subtropical parts of the eastern Transvaal passing through Zululand and Natal and even extends further into the eastern Cape in the Stockenstroom district. The greatest variation seen in the flower is in the length of the column of the corolla and the size of the terminal 5-winged knob which may be reduced to almost pin-head size.

subsp. **verruculosa** *R.A.Dyer* in Flora S.Afr. 27,4 (1980) [Fig.84A]
Ceropegia distincta subsp. *verruculosa* has minutely warty or rough stems with small hard pimples: the terminal portion of the corolla is narrowly cage-like and not knob-shaped: the roots are somewhat fleshy. The distribution is near Burgersfort in the dry northern Transvaal. It was first described and illustrated in Flowering Plants of Africa vol.44, Plate 1735 (1977). Dan le Roux, who has a nursery of succulent plants at P.P.Rust, northern Transvaal, assures me

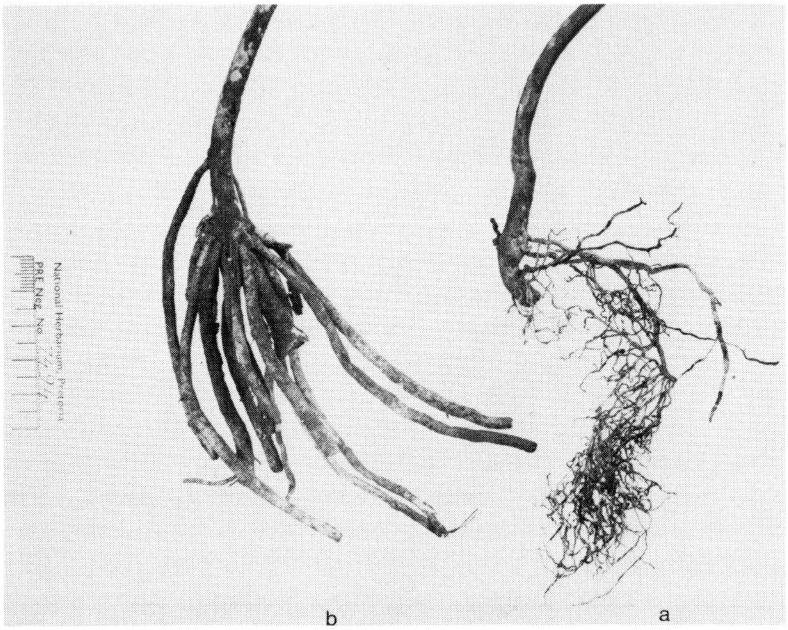

Figure 84B. *Ceropegia distincta:* a, root system of subspecies *haygarthii,* b, root system of subspecies *verruculosa* (specimens from *D.Roux*).

that the root-systems of the two subspecies have remained constant throughout the many years he has had the plants in cultivation.

It would seem from notes and drawings of material from Tropical Africa at Kew kindly prepared for me by David Field, that I have oversimplified the position, but at this juncture I cannot suggest an improvement.

16. **Ceropegia lugardiae** *N.E.Br.* in Gdnrs' Chron.30:302 (1901); R.A.Dyer in Flora S.Afr. 27,4:57 (1980) [Fig.85]

Perennial herb producing a cluster of fleshy roots. *Stems* perennial or produced annually from rootstock, slightly succulent, twining or rambling up to 2 m or

Figure 85. *Ceropegia lugardiae:* a, habit of plant from Okavango, SW Africa, x ⅔; b, flower, x 1 (*B.de Winter* 4853); c, (*Story* 5264, near Nama Pan, SW Africa).

more in length, sparsely branched. *Leaves* slightly fleshy, usually about 30 mm long, up to 20 mm broad. *Flowers* 3-6 in cymes lateral at the nodes, developing successively. *Corolla* 25-35 mm long; tube globose-inflated at base, abruptly up-curved above inflation, cylindric, then dilated into a funnel-shaped mouth, white at base outside, purple-spotted above on light green inflation and with sulphur-yellow mouth; lobes 10-20 mm long, connivent-erect, connate at tips, margins reflexed, purple-veined.

Ceropegia lugardae is named after the wife of Major E.J.Lugard who made two expeditions to Ngamiland in its early history. Mrs Lugard had a love of plants and collected and painted them with considerable artistic talent. The species was described by N.E.Brown in 1901. Huber regarded it as a form of *C.distincta* in his revision of the genus in 1958, but I do not agree with this arrangement and thus have restored its specific rank. It has a wide distribution in Tropical Africa and is not uncommon in South West Africa and Botswana where Mrs Lugard collected it on the Kwebe Hills.

17. **Ceropegia purpurascens** *K.Schum.* in Bot.Jb.17:152 (1893); R.A.Dyer in Flora S.Afr. 27,4:57 (1980)

Perennial herb producing a cluster of fleshy roots. *Stems* possibly arising annually, twining, up to a few metres long, with long internodes and with or without hairs. *Leaves* stalked, up to 70 mm long but usually much less, 20-30 mm broad. *Flowers* several – many in dense cymes, developed successively. *Corolla* ±20 mm long, with inflated base, 4-5 mm diameter bent at right-angles above, cylindric and dilated to ±7 mm at throat, green, mainly purplish within; lobes ± equal in length to tube slender, united at tips with reflexed margins with long fine purple hairs.

Ceropegia purpurascens has purplish flowers. The type specimen was collected near Pungo Andongo in Angola by Mechow and described by K.Schumann. The voucher specimen on which the present record is made was collected by Killick & Leistner 3267, in the Caprivi Strip and kindly identified by David Field at Kew. The species was also recorded from Botswana, from which source it was given the abandoned name *Ceropegia kwebensis,* based on specimens collected by Mrs E.J.Lugard in the same area and about the same time as she collected *C.lugardae,* dealt with under the previous number.

18. **Ceropegia cycniflora** *R.A.Dyer* in Bothalia 12,3:444 (1978); in Flora S. Afr. 27,4:57 (1980) [Fig.86]

Perennial with a cluster of succulent roots. *Stem* single, produced annually from rootstock up to 3 m tall, climbing and slightly twining, somewhat quad-

Figure 86. *Ceropegia cycniflora*, the only illustration of this rare species from Muden, Natal, x 2 (*W.E.Cronwright* 26).

rangular and more herbaceous than succulent, moderately hairy. *Leaves* shortly stalked, up to about 10 mm or more long and more or less heart-shaped at base and tapering. *Flowers* developed successively from small stalked groups, *Corolla* 30-35(40) mm long, lobed to about halfway, mainly light green; tube with a slightly elongated basal inflation ±6 mm diameter narrowed gradually into a short waist and then expanding to mouth about 6 mm diameter; lobes slightly spreading from the tube with infolded margins, furnished with long, clavate, vibratile caducous hairs, united at the tips.

Ceropegia cycniflora has a flower whose shape gave me the impression of a crane's head and prompted the specific name. It must be relatively rare because only W.E.Cronwright has recorded it and then in 1936 in the Muden Valley, Natal. Unfortunately interest in the genus was at a low ebb at the time with the result that there was no follow up of the discovery when it could have been rewarding. Perhaps this note will inspire an enthusiast to rediscover the species in its wild state near Muden on a steep woody hillside of the Mooi River valley. Its flowering period terminates about the middle of March.

19. **Ceropegia nilotica** *Kotschy* in Sitzungsb.Math: Nat.Akad.51:356 (1865); R.A.Dyer in Flora S.Afr. 27,4:59 (1980) [Fig.87]

Perennial with cluster of fleshy roots. *Stem* single, developed annually, somewhat fleshy, twining, 1 m or much longer. *Leaves* fleshy, stalked, up to 80 mm long, 60 mm broad. *Flowers* several, developing successively, lateral at nodes. *Corolla* 35-50 mm long; tube 25-35 mm long, inflated at base with slight con-

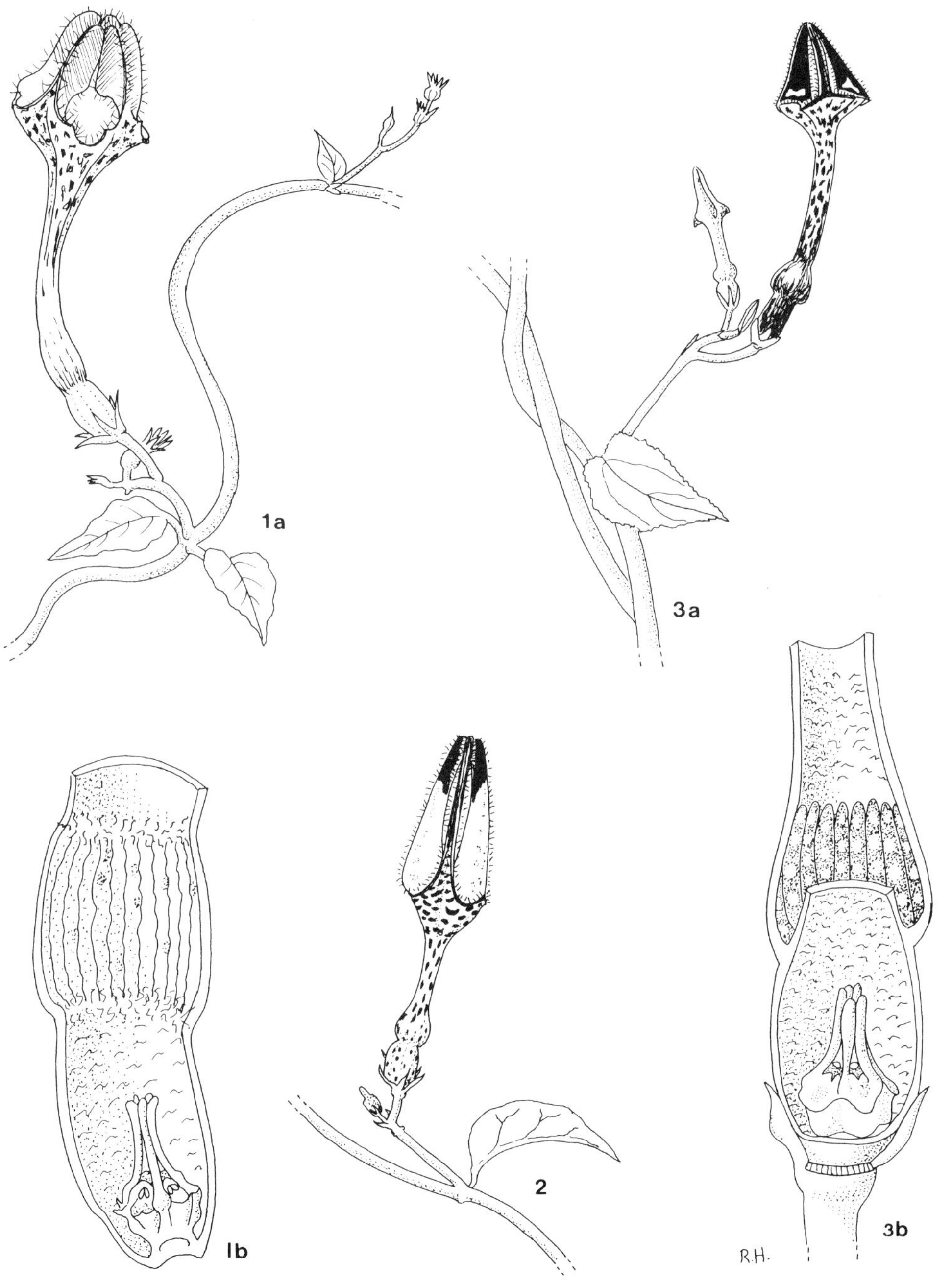

Figure 87. *Ceropegia nilotica:* 1, form previously described under the name *C.grandis,* 1a, twig, x ¾; 1b, longitudinal section of base of corolla showing constriction and corona, x 4½ (adapted from Flower. Pl. Afr. t.1113). 2, specimen very similar to typical form of species, x ¾ (*De Winter & Marais* 4560). 3, form described under the name *C.plicata,* 3a, twig, x ¾; 3b, longitudinal section of base of tube showing extension of constriction within and exposing corona, x 4½ (adapted mainly from Flower. Pl. Afr. t.675) (Fig.12 in Flora S.Afr. 27,4:(1979)).

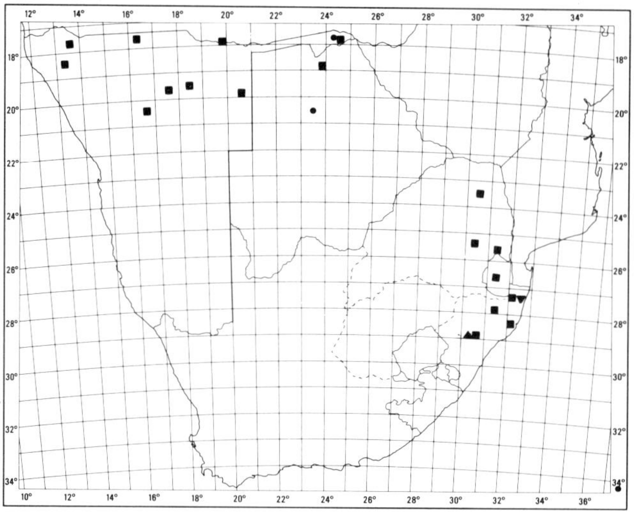

● *Ceropegia purpurascens* (17)
▲ *Ceropegia cycniflora* (18)
■ *Ceropegia nilotica* (19)
▼ *Ceropegia arenaria* (20)

striction near middle or top of inflation or both, with the middle coinciding with a thickening or collar from the inner wall, spreading at mouth and sinuses, mainly whitish with mauve markings; lobes united, varying greatly in shape, forming a short to long pyramidal cage, blackish-purple and hairy within, marked with yellow spots near base.

Ceropegia nilotica as the name implies originated from the valley of the Nile River in north Africa. Several variations have been included under the name with the result that the distribution as accepted as being almost throughout tropical Africa, southwards into South West Africa, Transvaal, Moçambique, Swaziland, Zululand and Natal. No two workers are likely to agree entirely on the limits of the species. While I have adopted a broad view in the synonymy, I am inclined to give *C.plicata* Bruce some distinctive status even if only as 'form'. Anyone wishing to revise the group will need to consult a mass of literature and study a good range of material, preferably living.

20. **Ceropegia arenaria** *R.A.Dyer* in Bothalia 12,3:444 (1978); Flora S.Afr. 27,4:59 (1980) [Fig.88]

Perennial with tuft of subsucculent or fibrous roots. *Stems* prostrate or rambling (not found twining) sometimes rooting at nodes, succulent, ±3,5 mm thick, slightly mottled. *Leaves* sometimes appearing obsolete, shortly stalked, up to 40 mm long and 25 mm broad. *Flowers* 2 – several, developed successively lateral at nodes. *Corolla* up to about 60 mm long; tube about 35 mm long, with basal inflation ±14 mm long and slightly narrowed upwards, then abruptly

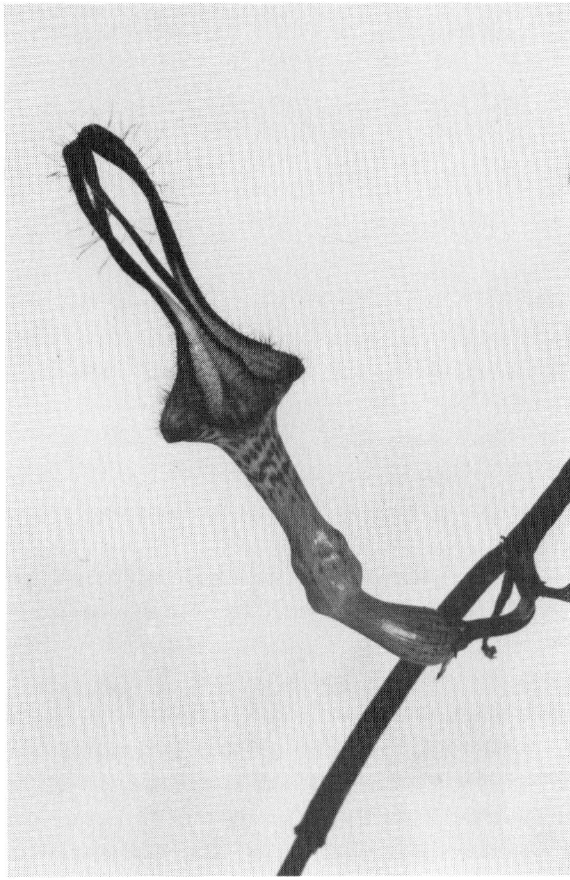

Figure 88. *Ceropegia arenaria:* flower from type, x 1½ (*R.Strey* 5031, Zululand, coastal bush).

spreading into a subglobose inflation 6-7 mm diameter, without an internal annulus, cylindric above, widening to ±15 mm at throat; lobes 20-25 mm long, united at tips, slightly converging at middle, margins strongly recurved, with long spreading hairs towards base and long purple, clavate, vibratile hairs towards apex, purplish.

Ceropegia arenaria whose natural habitat is in sandy soil, is known from only two collections, both in Zululand near the coast between Lake Mpangazi and Sodwana Bay in sand-dune vegetation. Stems were found to be rambling in the undergrowth and none was actually climbing or twining. One collection was of a plant with developed leaves and the other without, but there was no apparent difference in the floral structure. Obviously further field observations are necessary in order to obtain a complete picture of the species. Its closest affinity is with the tropical species *Ceropegia denticulata* var. *brownii* established by P.R.O.Bally of Kenya.

Figure 89. *Ceropegia stenantha:* the long beak and spreading sinuses at the base of the corolla lobes are noteworthy features, x 1½ (*B.de Winter & W.Marais* 4736, Okavango, SW Africa).

21. **Ceropegia stenantha** *K.Schum.* in Bot.Jb.17:152 (1893); R.A.Dyer in Flora S.Afr. 27,4:61 (1980) [Fig.89]

Perennial producing a cluster of long, fleshy roots. *Stems* 1-several, produced annually, climbing and twining, 1-few metres long, without hairs. *Leaves* somewhat fleshy, shortly stalked, slender, up to 80(100) mm long. *Flowers* 3-6 together, lateral at the nodes. *Corolla* straight or slightly curved, 25-30 mm long, greenish yellow, lobed to about halfway, tube 20 mm long, oblong-inflated at base, 2-3 mm diameter, narrowed above, not dilated at mouth but with five spreading points; lobes ±15(20) mm long, slender, united at tips forming an elongated cage, occasionally becoming partly free.

Ceropegia stenantha, the species with the 'slender flower', has a wide distribution in tropical Africa but is rare in the northern tropical parts of southern Africa. It was brought into cultivation at the BRI in 1956 by an expedition to the Okavango by B.de Winter and W.Marais, 4736, and was grown successfully for one season when the accompanying photographs were taken. C.J.Ward, 3119, collected it in northern Zululand in 1957 and it has not been recorded again in our area.

Without a knowledge of the rootsystem of a cluster of fleshy roots, this species could be mistaken at first sight for *C.floribunda,* 34a, which, however, grows from a tuber.

22. **Ceropegia radicans** *Schltr.* in Bot.Jb.18 Beibl.45:12 (1894); R.A.Dyer in Flora S.Afr. 27,4:61 (1980) [Fig.90]

Perennial succulent, producing a cluster of cylindric roots and sometimes rooting at nodes. *Stems* prostrate, sparingly branched. *Leaves* fleshy, up to 40 mm long, 25 mm broad. *Flowers* 1-3, lateral at the nodes, produced in succession. *Corolla* 55-80 mm long; tube 40-55 mm long, with slightly 5-angled basal inflation 13-17 mm long and 5-7 mm diameter, contracted to 3 mm, then expanding gradually to the mouth of about 15 mm diameter, softly hairy within inflation, pale greenish white with purple dots and splotches, intensified towards mouth.

Two subspecies are distinguished thus:

Corolla-lobes forming an elongated pyramid up to 25 mm long to the united tips
subsp. **radicans**

Corolla-lobes obcordate-oblong, forming an obtuse cage-like structure about 15 mm long, slightly narrowed about the middle subsp. **smithii**

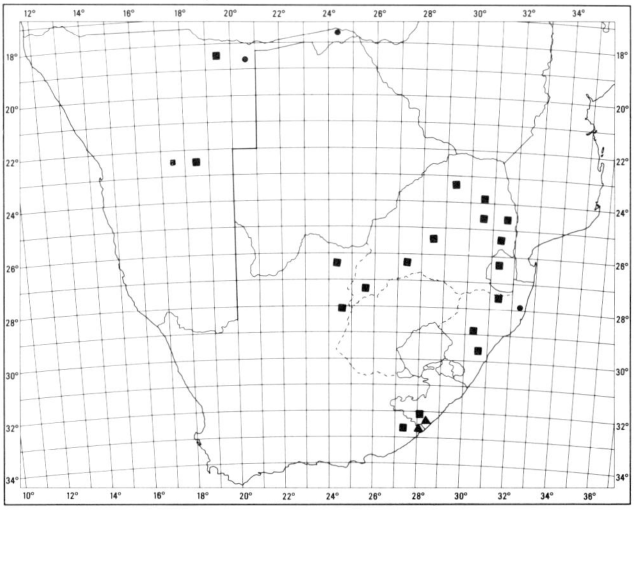

● *Ceropegia stenantha* (21)
▲ *Ceropegia radicans* (22)
■ *Ceropegia crassifolia* (23)

Figure 90. *Ceropegia radicans:* a, subsp. *radicans,* normally prostrate, x ¾; b, subsp. *smithii,* x ¾; valleys of Ciskei, eastern Cape (after BRI records).

subsp. **radicans** *R.A.Dyer* in Flora S.Afr. 27,4:62 (1980) [Fig.90a]
Corolla-lobes connivent-erect, up to 25 mm long, connate at tips, cage-like, purple-brown within base, with broad transverse whitish band near base, then dark-purple striped and bright green in upper half, thinly hairy on inner surface and with purple, clavate-vibratile hairs on the margins towards the tips.

Ceropegia radicans subsp. *radicans* so named because of its prostrate habit and consequent rooting at the nodes was described by R.Schlechter in an account of collections of Asclepiadaceae sent to him by H.G.Flanagan and E.E.Galpin, mainly from the eastern Cape Province. The record for *C.radicans* was by Flanagan from scrub-bush near Komga at an altitude of 580 m where it flowered in January 1890. At the time it was the largest flower known to Schlechter in the genus. It has been recorded several times in this border area but nowhere further afield.

subsp. **smithii** *(M.R.Henderson) R.A.Dyer* in Flora S.Afr. 27,4:62 (1980)
[Fig.90b]

Corolla-lobes forming a cage about 15 mm long, with flat top and slightly narrowed about the middle; colouring and hairiness as for subsp. *radicans.*

Ceropegia radicans subsp. *smithii* is named after G.G.Smith of East London, Chairman for many years of the Board of Trustees of the local Museum. It was given specific status at first by Mr M.R.Henderson based on a specimen collected by G.G.Smith in the Kwelegha River valley about 48 km north-east of East London, where it grew under bushes on south to south-east facing slopes. It has not been located elsewhere and is in an adjacent valley to subsp. *radicans.* An attractive feature of subsp. *smithii* on occasions is the marbled effect of the leaves caused by light and dark green marking along the main veins.

23. **Ceropegia crassifolia** *Schltr.* in J.Bot.33:273 (1895); R.A.Dyer in Flora S.Afr. 27,4:62 (1980) [Plate 15, Fig.91, 91A]

Perennial producing a cluster of fleshy roots. *Stem* succulent, unbranched or sparingly branched, from dwarf to 1 m or more tall, twining above, rarely scrambling. *Leaves* fleshy, sometimes crisped on margin, up to 80 mm long, 45 mm broad on first growth soon reduced on upper growth. *Flowers* 3-7 lateral at nodes, opening in succession or sometimes two open at same time. *Corolla* 25-50 mm long; tube about two-thirds total length, slightly curved to nearly straight, globose-inflated at base up to 9 mm diameter, subcylindric and expanded to mouth about 9 mm wide, pale green with purplish markings and striped within; lobes united at tips, opening into a rounded cage 10-15 mm diameter, with inner surface finely reticulate with purple-brown, with margins recurved and long purple hairs on basal part and inner keel.

Ceropegia crassifolia, with fleshy leaves, was described in one of a series of contributions on the Asclepiadaceae of Southern Africa by R.Schlechter. The type specimen was collected in the scrub-bush near King William's Town in the eastern Cape by T.R.Sim 312 in 1893. Evidently Sim had placed it in boiling water preparatory to pressing and drying it, which rendered it unsuitable for Schlechter to make a dissection for description. This deficiency was made good by N.E.Brown and other workers under various names now in the synonymy of *C.crassifolia* for example *C.crispata* and *C.thorncroftii* N.E.Br. and *C.tuberculata* Dinter.

The distribution range is wide, from the northern parts of S.W.Africa into northern, western, and the warmer parts of the eastern Transvaal, into Swaziland, Zululand, Natal and eastern Cape near Uitenhage. Over this wide area the colour markings and other characters vary somewhat but not regularly enough to justify the subdivision of the species.

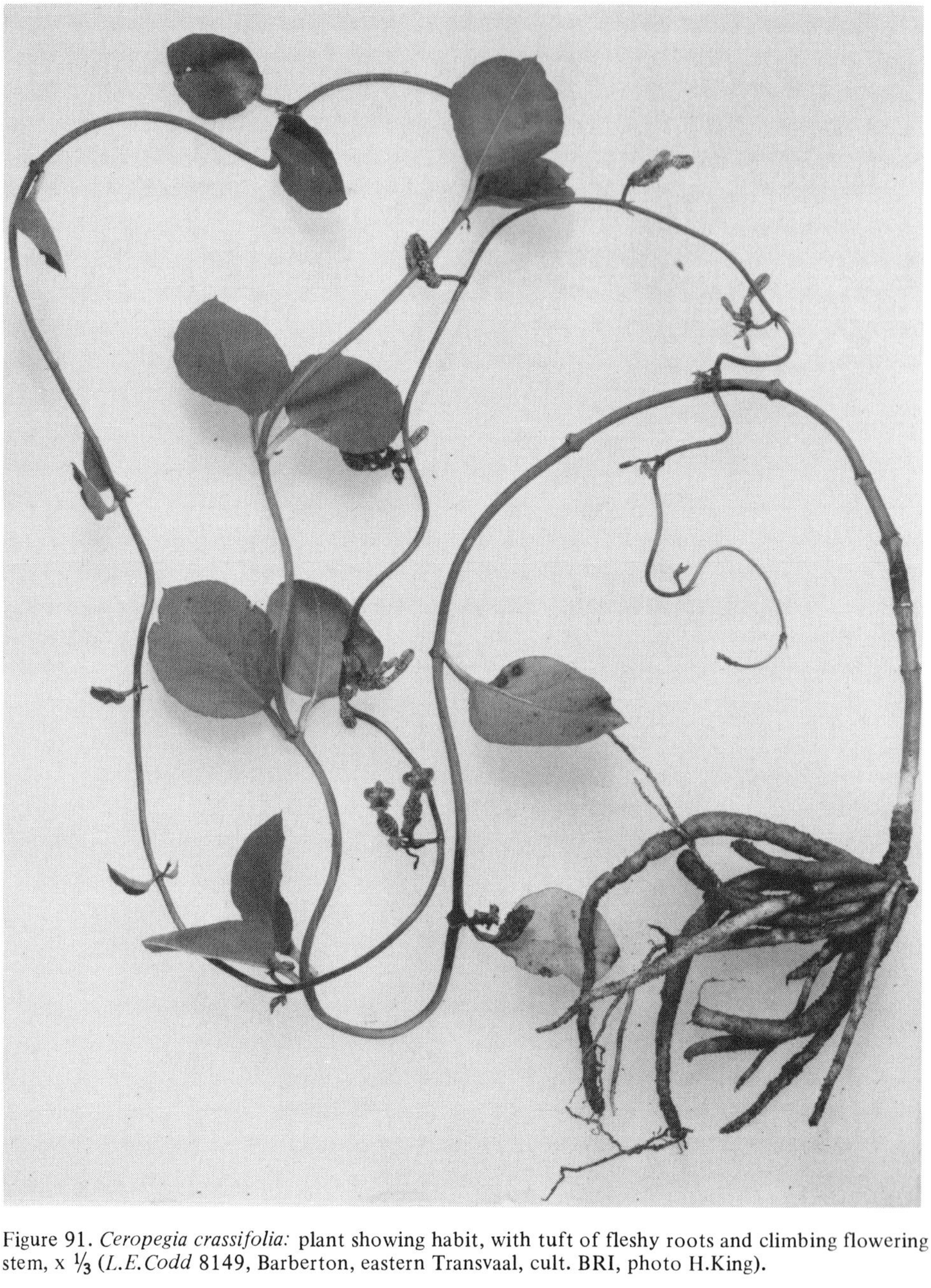

Figure 91. *Ceropegia crassifolia:* plant showing habit, with tuft of fleshy roots and climbing flowering stem, x ⅓ (*L.E.Codd* 8149, Barberton, eastern Transvaal, cult. BRI, photo H.King).

Figure 91A. *Ceropegia crassifolia:* a, plant becoming very floriferous in cultivation, x ¼; b, x ½ (*L.E.Codd* 8149, Barberton, eastern Transvaal, cult. BRI, photo H.King).

24. **Ceropegia carnosa** *E.Mey.*, Comm.193 (1837); R.A.Dyer in Flora S.Afr. 27,4:63 (1980) [Plate 16, Fig.92]

Perennial herb with rootstock producing a cluster of fleshy roots, some or all of which may produce swollen tuberous storage organs towards their lower extremity. *Stems* climbing-twining, up to about 2 m long and 20 mm broad. *Corolla* up to 20 mm long; tube up to 15 mm long, cylindric, expanded to about 7 mm at mouth, with slightly spreading points; lobes united at apex, about 6 mm long, with reflexed margins, forming a cage-like body broader than the mouth of the tube, finely ciliate with long white hairs.

Ceropegia carnosa received its name because of its somewhat fleshy habit. Although it was discovered by J.F.Drège about 1829, the species has not been

Figure 92. *Ceropegia carnosa,* flower, x 2¼ (*L.E.Codd* 9634, Zululand, cult. BRI).

illustrated previously in any scientific publication. Auriol Batten thus broke new ground when she painted the accompanying illustration. In addition she made the discovery of the deeply seated large tuberous storage organs on some (or all?) of the roots. Hitherto the roots had been assumed to be of nearly uniform thickness throughout. It is not known whether any other species shares this character.

The distribution of *C.carnosa* extends in association with scrub-bush from the Uitenhage area of the eastern Cape to Transkei and Natal, from where it passes into Zululand and Swaziland.

The flowers of *C.carnosa* resemble those of the more robust *C.racemosa* subsp. *setifera* in which the corolla-cage is narrower and longer in proportion as compared with the shorter and broader cage of *C.carnosa.* The structure of the corona, more particularly of the outer series, supplies a final answer. In *C.carnosa* the outer lobes of the corona are erect or curved and deeply 2-toothed or 2-lobuled, sometimes giving the impression of 10 equal teeth as opposed to those of *C.racemosa* which appear as shoulders to the inner corona series with which they are confluent.

Plate 16. *Ceropegia carnosa* (species 24); a. rootstock with somewhat fleshy roots and remarkable tuberous storage roots, x 1; b. flowering stems, x 1; c. longitudinal section of bottom half of corolla, x 4; d. corona with outer lobes deeply bilobuled, x 8; e. pollinia with short caudicles (attachments) to shortly winged carrier, x ±40 (*Auriol Batten* 323, coll. near type locality in Fish River Valley, eastern Cape)

a
b
c
d
e
Auriol Batten

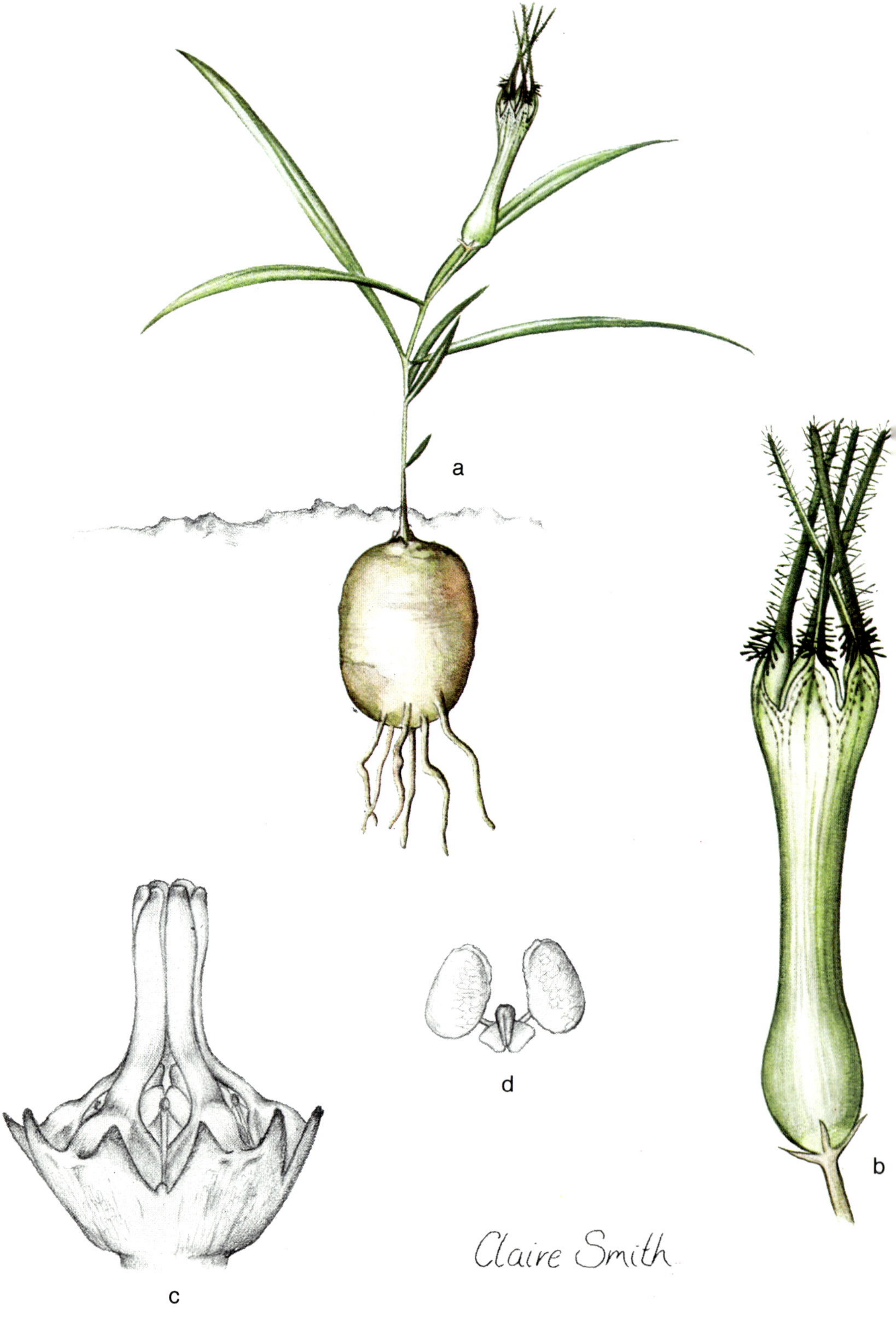
a
b
c
d
Claire Smith

25. **Ceropegia racemosa** *(N.E.Br.)* subsp. **setifera** *(Schltr.) Huber* in Mems Soc. broteriana 12:93 (1958); R.A.Dyer in Flora S.Afr. 27,4:63 (1980)

[Fig.93, 93A]

Perennial herb producing a cluster of elongated fleshy roots. *Stem* twining, up to 2 m or more tall, rarely dwarf and less than 200 mm, thinly hairy or without hairs. *Leaves* stalked, blade 25-80 mm long, 12-45 mm broad, variably hairy. *Flowers* 2-several, developing successively lateral at the nodes. *Corolla* about 30 mm long, greenish to yellow, variably purplish marked and darker on the margin of lobes; tube about 22 mm long, sometimes slightly curved from slightly inflated base, widening to about 10 mm at mouth, and spreading into 5 points in bud; lobes united at tips into a roundish cage, broader than the mouth of the tube, with few long, pale purple to white hairs along the keel within.

Ceropegia racemosa subsp. *setifera,* so named because of the character of the inflorescence and corona-lobes is a subspecies of a very variable species with a wide distribution in tropical Africa and with the subspecies entering Southern Africa through the Transvaal and fading out in southern Natal. Certain subspecies and varieties have been recognised at different times and it may be over-simplifying the position to recognise only one subspecies here. This has resulted in a dwarf, alpine plant from The Downs in northern Transvaal being included too. This latter is represented by a collection made by Alan Percy-Lancaster under his number 200 and another by E.A.Pienaar. The response of this dwarf mountain form to cultivation in the lower and warmer

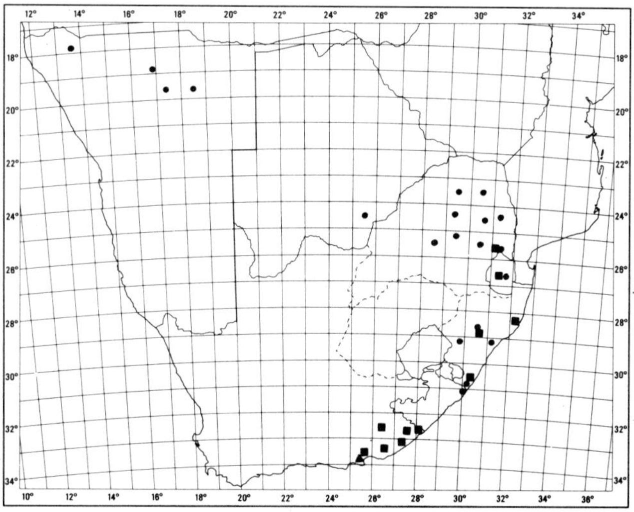

■ *Ceropegia carnosa* (24)
● *Ceropegia racemosa* subsp. *setifera* (25)
▲ *Ceropegia dubia* (26)

Plate 17. *Ceropegia dinteri* (species 29); a. young flowering plants, x 1; b.flower, x 3; c. corona, x 10; d. pollinia attached to winged carrier, x 60 (painting by Claire Smith, BRI; Coll. P.Bruyns)

Figure 93. *Ceropegia racemosa* subsp. *setifera:* a, habit, x ½; b, flower, x 1½ (*L.E.Codd* 9552, near Barberton, eastern Transvaal, cult. BRI).

Figure 93A. *Ceropegia racemosa* subsp. *setifera:* dwarf form, ±100 m high, not elongating or twining in the wild state but producing climbing stem in cult. BRI (*Pienaar & Lafon,* The Downs Mt. northern Transvaal).

environment of Pretoria at BRI was to produce climbing stems, as seen in Figure 93A.

Some difficulty arises occasionally in distinguishing between *Ceropegia racemosa* subsp. *setifera* and its smaller relative *Ceropegia carnosa* with somewhat succulent leaves. The most reliable distinctions are in the corolla-cage and corona. In *C.carnosa* the outer corona is erect and the lobes deeply 2-toothed or bilobuled, whereas in *C.racemosa* subsp. *setifera* the outer lobes appear as shoulders to the inner corona lobes with which they are confluent.

26. **Ceropegia dubia** *R.A.Dyer* in Bothalia 12,4: (1979); in Flora S.Afr. 27,4: 64 (1980) [Fig.94]

Perennial herb producing a cluster of succulent roots. *Stems* slender, climbing, branched, up to several metres in cultivation, hairy. *Leaves* shortly stalked, 15-25 mm long, 10-15 mm broad, hairy. *Flowers* 2 in succession lateral at the

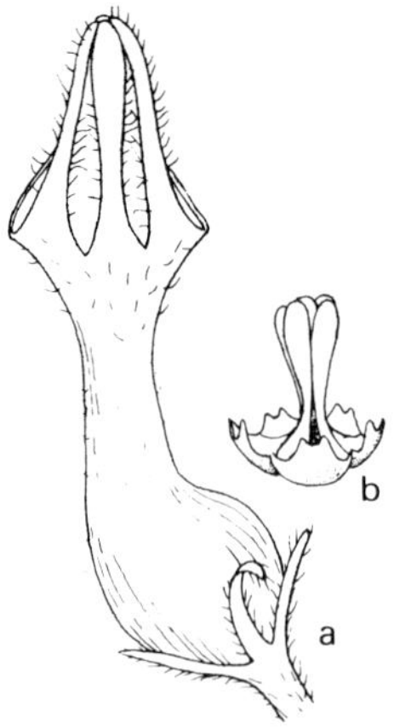

Figure 94. *Ceropegia dubia:* a, drawing of type flower, x 2; b, corona with 5 chute-like outer lobes and inner lobes subspathulate, erect, somewhat fleshy, x 5 (*R.Bayliss* 7280, between Coega and Uitenhage, eastern Cape).

nodes. *Corolla* 22-25 mm long, slightly curved; tube ±15 mm long with basal globose inflation about 5 mm diameter contracted above and then somewhat abruptly widened to a trumpet-shaped mouth; lobes united at tips forming a cage 6-7 mm long, margins folded back, with long hairs on margin.

Ceropegia dubia leaves me in doubt as to its nearest relatives and derivation. I find it hard to reconcile so late a discovery as 1976 of this species from such a well-botanised area as the neighbourhood of Uitenhage in the eastern Cape. It was recorded from there by the most reliable collector, Roy Bayliss, who reported that he had subsequently had it in cultivation in Grahamstown where it flowered and fruited copiously for over 12 months. Its tuft of fleshy roots and no tuber distinguish it from the tuberous rooted species *C.africana, C.cancellata* and *C.pachystelma* which it resembles in habit.

27. **Ceropegia pygmaea** *Schinz* in Verh.bot.Ver.Brandenb.30:265 (1888); R.A. Dyer in Flora S.Afr. 27,4:64 (1980) [Fig.95]

Perennial dwarf herb with tuber. *Tuber* roundish, 25-50 mm diameter. *Stems* annual, 1-3, unbranched or sparingly branched, erect, 100-120 mm tall, variably hairy. *Leaves* shortly stalked; blade somewhat slender, 25-100 mm long. *Flowers* 1-3 on short stalk lateral at the nodes, opening in succession. *Corolla* 30-50 mm long, very variable in size, hairy outside, without hair within, dark purple, inflated near base with age, subcylindric above and sharply bent, slightly expanding to throat; lobes 2-3 mm long, free, somewhat incurved, with recurved margins.

Ceropegia pygmaea, a dwarf, like the bushmen who feed on the tubers, shows a great range in the size and shape of the corolla. This variation gave rise to the additional names *Ceropegia gymnopoda* Schltr. and *C.pumila* N.E.

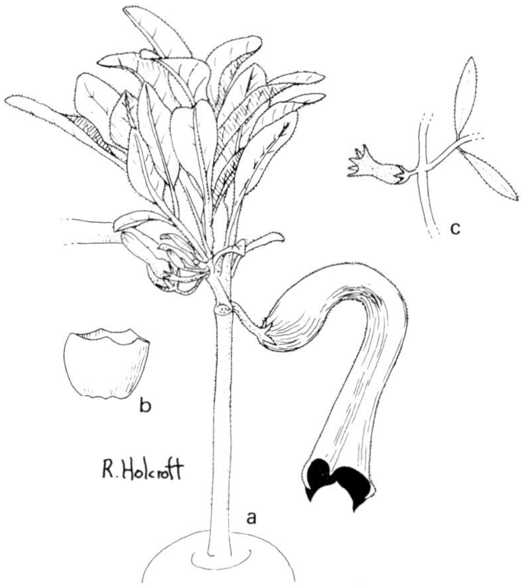

Figure 95. *Ceropegia pygmaea:* a, typical form of the species, x ½; b, corona of typical form, x 3; c, form once given the specific name *pumila* and indicating the wide range of variability within the species (after *Dinter & N.E.Brown*).

Br. and later to the treatment of the latter dwarf form as *C.pygmaea* var. *pumila.* On the assumption that the variability is linked with local environment one might justify the variety. But the distribution is from Angola and Rhodesia to northern parts of SWA to Botswana, northern Cape and northern Transvaal and to eastern Transvaal lowveld near Phalaborwa, with no marked isolation of forms.

The tubers are eagerly sought after by the aborigines as food.

28. **Ceropegia mafekingensis** *(N.E.Br.) R.A.Dyer* in Bothalia 12:256 (1977); Flora S.Afr. 27,4:65 (1980) [Fig.96]

Perennial dwarf herb with tuber. *Tuber* about 80 mm diameter, somewhat depressed. *Stems* 1-2, annual branching from near base and appearing tufted, 15-50 mm high, hairy. *Leaves* shortly stalked, 7-30 mm long, up to 7 mm broad, with margins somewhat folded upwards, more or less undulate,hairy on under-surface, without hairs on upper surface. *Flowers* 10-20 together, compact, several opening together. *Corolla* 10-12 mm long, divided to more than half-way; tube 4-6 mm long, 3-3,5 mm diameter, obtusely 5-angled, slightly narrowed to mouth, mottled dark purple, yellow and purple-punctate within; lobes 6-7 mm long, free at tips, spreading with age, about 2 mm broad, eared at base, margins folded back forming keel down face, warty, blackish.

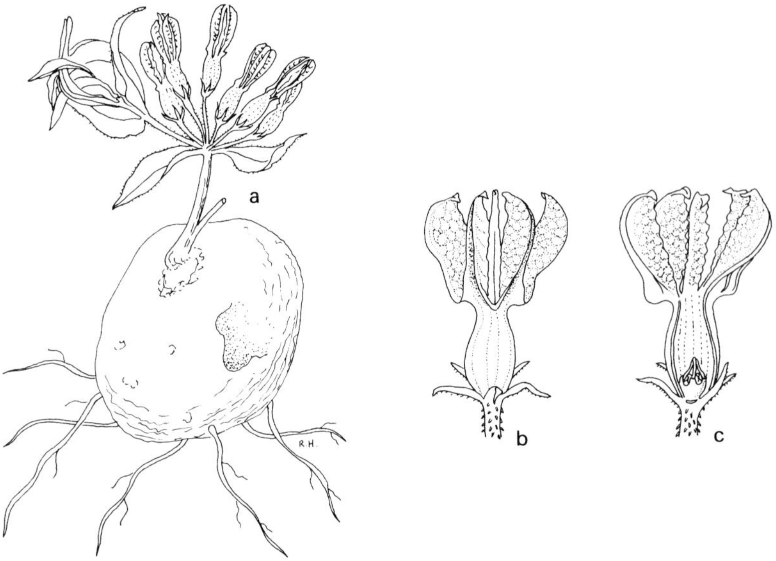

Figure 96. *Ceropegia mafekingensis:* a, form given the name *C.patriciae,* from north of Pretoria, x ½, the typical form from Mafeking in Kalahari sandveld may have the tuber about 10 cm underground; b, flower, x 1¼; c, longitudinal section of flower showing corona, x 1¼ (after *Prof. Rauh* 12369, with his kind permission).

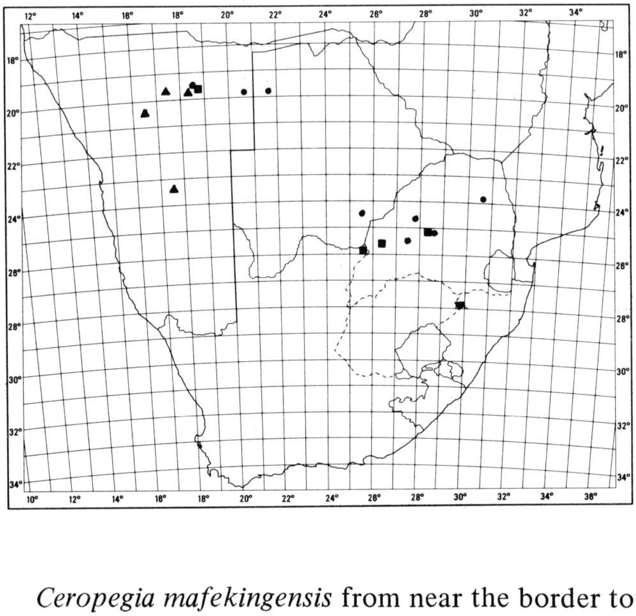

● *Ceropegia pygmaea* (27)
■ *Ceropegia mafekingensis* (28)
▲ *Ceropegia dinteri* (29)
▼ *Ceropegia antennifera* (30)

Ceropegia mafekingensis from near the border town of Mafeking on the boundary between Botswana and western Transvaal, has had a checkered career. It was for many years left in the genus *Brachystelma* and only relatively recently transferred to *Ceropegia.* Its history is sketched in Bothalia 12: 256 (1977). Of recent years it has been traced rather sporadically from SWA to the north-eastern Cape, western and northern Transvaal. Like the majority of tuberous rooted species, the tubers are relished by the aborigines, particularly Bushmen, and it is inevitable that this has had an effect on the distribution frequency of the species.

29. **Ceropegia dinteri** *Schltr.* in Bot.Jb.51:53 (1913); R.A.Dyer in Flora S. Afr. 27,4:65 (1980) [Plate 17, Fig.97]

Perennial herb with tuber. *Tuber* variable in shape and size from flattened up to 120 mm broad and 50-60 mm thick to turnip or top-shaped and flowering at 20 mm diameter. *Stem* annual, herbaceous, solitary, erect, unbranched, without hairs, from 150-350 mm tall. *Leaves* sessile, narrowly linear, 120-160 mm long, about 4 mm broad, tapering to tip and base. *Flowers* solitary

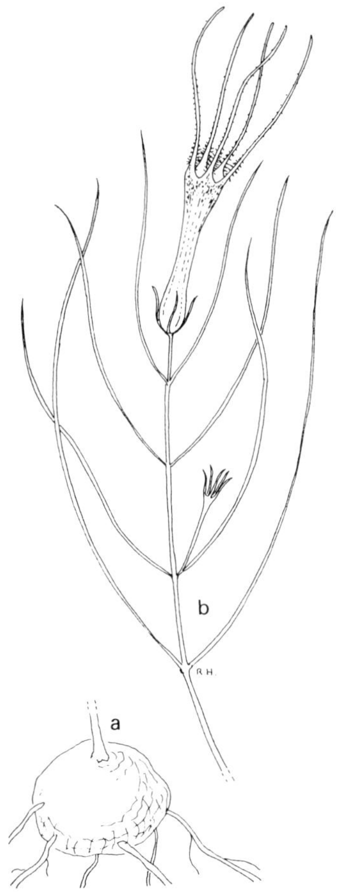

Figure 97. *Ceropegia dinteri:* a, tuber and b, annual flowering stem with apparently disproportionately large flower, x ½ (after *Dinter* 2529).

on short ascending stalk. *Corolla* erect, variable in size from about 40-85 mm long, with tube usually somewhat longer than the lobes; tube slender with base slightly inflated about 6 mm diameter and throat slightly wider, from a greenish base to yellowish white above; lobes free at the tips, variable in length in proportion to the length of the tube, from a reddish brown triangular base abruptly contracted to linear above, brown, with long thick subcylindric vibratile hairs at the base of the contracted portion and with long thin fixed hairs on the linear portion above.

Ceropegia dinteri, named after the famous German collector in SWA, Kurt Dinter, is restricted to SWA where it is distributed under various conditions in the districts of Autjo, Tsumeb, Grootfontein, Otjiwarongo and Rehoboth and probably over the borders of neighbouring districts. The corolla-lobes tend to cross over each other and do not spread regularly from the tube as one might expect. The two contrasting shaped hairs on the lobes are unusual in a genus noted for its wealth of nature's wonders.

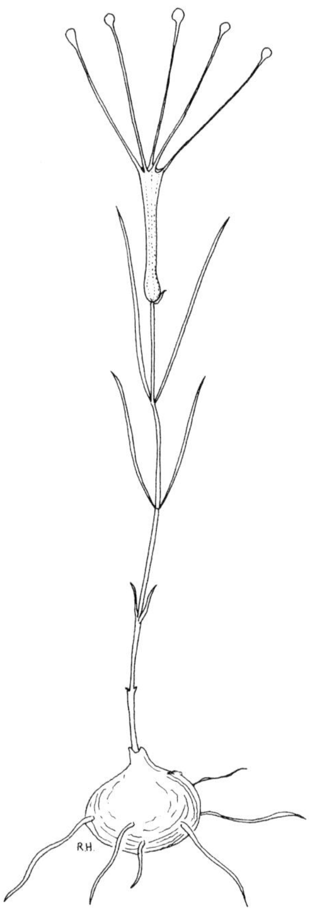

Figure 98. *Ceropegia antennifera:* whole plant showing yet another remarkable variation of the corolla structure, x ½ (after *Schlechter's* type 3426, from near Newcastle, Natal).

The young and enthusiastic collector of Ceropegias, Peter Bruyns, has made it possible to illustrate this relatively little-known species in colour. He gathered tubers near Outjo, some of which he sent to E.A.Pienaar in Pretoria for cultivation and it is one of these which is the subject of the present painting. From the field notes of Peter Bruyns it is clear that the average size of tubers and plants near Outjo, under various conditions of the habitat, are considerably smaller than those originally collected by Kurt Dinter.

30. **Ceropegia antennifera** *Schltr.* in Bot.Jb.20: Beibl.51:46 (1895); R.A.Dyer in Flora S.Afr. 27,4:66 (1980) [Fig.98]

Perennial herb with tuber. *Tuber* 25-30 mm diameter, depressed. *Stem* solitary, annual, herbaceous, erect, 120-200 mm high. *Leaves* in 5-6 pairs, very slender, up to about 300 mm long, 1 mm broad. *Flowers* solitary, lateral at nodes or subterminal, erect. *Corolla* 75-95 mm long; tube about 35 mm long, with inflated base about 4 mm diameter, narrowed to 2 mm and 3 mm at throat, without hair, greenish-striped with purple-brown; lobes free, erectly spreading, slender with expanded tips, 50-60 mm long, bearded with purple hairs.

Ceropegia antennifera was claimed by Schlechter the collector and author of the type specimen and species to have the most unusual corolla-lobes he had met within the genus, resembling as they do, the antennae of certain insects. The rarity of the species may be judged by the fact that no further record has been added to that of Schlechter in grassveld near Newcastle, Natal, over 80 years ago. South African botany is fortunate in that a duplicate of the type is housed in the Bolus Herbarium of the University of Cape Town.

31. **Ceropegia insignis** *R.A.Dyer* in Flower.Pl.S.Afr.23:5.902 (1943); in Flora S.Afr. 27,4:66 (1980) [Fig.99]

Perennial herb with tuber. *Tuber* about 45 mm diameter, somewhat depressed, to about 30 mm high. *Stem* single annually, 200-450 mm tall, unbranched, with several internodes 30-50 mm long, with reflexed hairs. *Leaves* slender, tapering to both ends, the middle ones 120-140 mm long. *Flowers* solitary, lateral at the nodes. *Corolla* about 120 mm long, mainly reddish brown; tube about 70 mm long, inflated with cup-shaped base, contracted to 4 mm diameter, and broadening to 15-17 mm diameter, at mouth; lobes about 50 mm long twisted in bud and the tips connate and twisted in open flower, margins reflexed near base.

Ceropegia insignis has a remarkable corolla in a genus rich in unusual forms. The long flower with the lobes twisted in bud has no close relative locally. Huber, l.c.131, placed it under the umbrella species *Ceropegia campanulata*

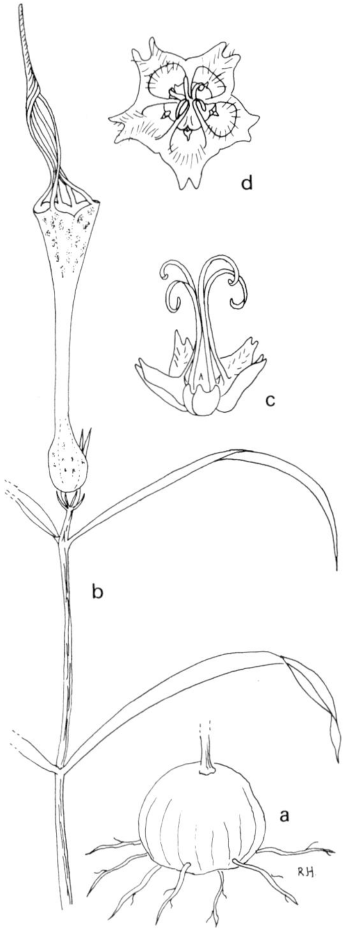

Figure 99. *Ceropegia insignis:* a, tuber, x ½; b, annual flowering stem with apparently disproportionately long flower, x ½; c, corona side view, x 3½; corolla face view, x 3½ (after *Louw* 812, Flower. Pl. Afr. t.902).

from Tropical Africa as var. *pulchella* but I do not accept this viewpoint and maintain its distinctive classification. The type was collected by W.J.Louw in 1940 in stony grassveld in the Marico district of the western Transvaal. A single plant was recorded some years later by Rev. Reinecke 7 km north of Kayaseput, also in the Marico district.

32. **Ceropegia stentiae** *E.A.Bruce* in Kew Bull.1936:490 (1936); R.A.Dyer in Flora S.Afr. 27,4:66 (1980) [Fig.100]

Perennial dwarf herb with tuber. *Tuber* about 30 mm broad, somewhat depressed, 20 mm high. *Stem* single, annually, up to about 12 cm tall sometimes sparingly branched, with internodes 5 mm long. *Leaves* very slender, 40-60 mm long, 2-3 mm broad. *Flowers* solitary, lateral at nodes. *Corolla* 45-

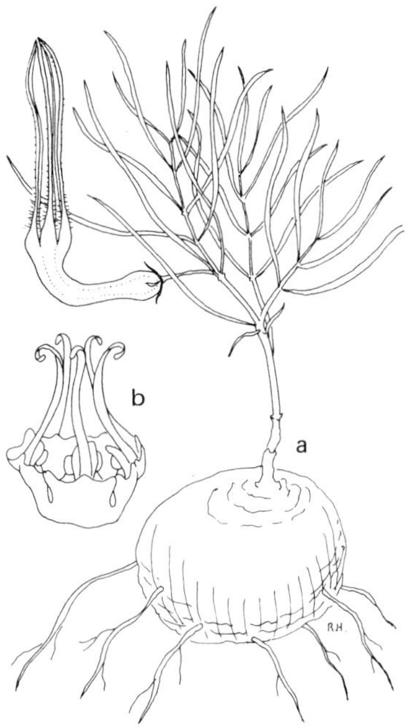

Figure 100. *Ceropegia stentiae:* a, habit sketch of plant with tuber and flower, x ½; b, corona, deeply cut opposite the pockets and with incumbent-erect-spreading inner lobes, x 4 (after *Galpin* M215, Waterberg, north of Pretoria).

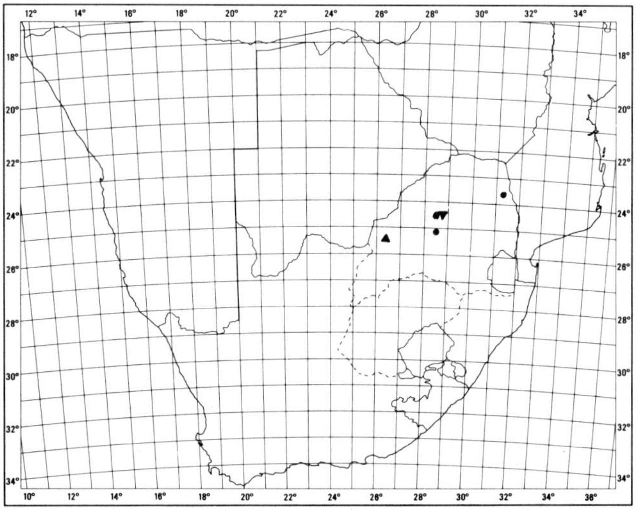

▲ *Ceropegia insignis* (31)
▼ *Ceropegia stentiae* (32)
● *Ceropegia turricula* (33)

70 mm long, tubular in lower half, spreading horizontally first then bent upright more or less at right-angles; tube 20-30 mm long, 3-4 mm diameter widening to 6-7 mm at throat, brownish purple with greenish stripes to base; lobes very slender, slightly broadened at the united tips, 25-40 mm long, with purple hairs on margin.

Ceropegia stentiae is named after Sidney Stent, a noted authority on South

African grasses in the early days of the Botanical Research Institute. She collected her specimen near Naboomspruit on the farm Mosdene owned by the Galpin family. E.E.Galpin collected the species in the same area of the northern Transvaal, Waterberg, and attributed the comparative rarity of the species, in some measure, to the edibility of the tubers which were keenly sought after by herdsmen. The slender corolla-lobes united at the tips, form long open cages with only the basal parts adorned with long cilia.

33. **Ceropegia turricula** *E.A.Bruce* in Kew Bull. 1936:491 (1937); R.A.Dyer in Flora S.Afr. 27,4:67 (1980) [Fig.101]

Perennial herb with tuber. *Tuber* 35-45 mm diameter, slightly broader than high. *Stem* single, annual, unbranched up to about 300 mm tall, with inter-

Figure 101. *Ceropegia turricula:* a, perennial underground stem with tuber, x ½; b, annual flowering stem with large flower, x ½; c, corona with two-toothed outer lobes and incumbent-erect inner lobes, x 5 (after Flower. Pl. Afr. 5.1045, northern Transvaal).

nodes 10-20 mm long. *Leaves* slender, up to 120 mm long, 5 mm broad. *Flowers* solitary. *Corolla* up to about 65 mm flat-topped or slightly tapered, 5-angled in bud; tube about 50 mm long, with inflated base about 7 mm diameter, narrowed to 3,5 mm and expanded to about 17,5 mm at mouth, pale green with purple-brown mottling outside, purple-ribbed within inflation, and with long purple hairs within throat and on base of lobes; lobes united at tips into a cage-like structure 17 mm long, 6 mm broad near top, slightly twisted at middle, margins becoming strongly recurved, with long thick hairs towards tips, inner face pale-green near base, transversely white banded followed by dark purple and green in upper half.

Ceropegia turricula has the corolla-lobes united at the tips to form a miniature turret above the tube. E.E.Galpin, regarded in his time as the prince of collectors and who gathered the type of this species near Naboomspruit in the northern Transvaal, recorded that he found only one specimen in six years of hunting and that the tubers appeared to be on the point of extinction because they were sought and eaten by the aborigines. Since then, however, a few isolated specimens have been collected on the Waterberg, near Pietersburg and in the eastern lowveld near Gravelotte, so it is not so near to extinction as Galpin feared.

34. **Ceropegia conrathii** *Schltr.* in Bot.Jb.38:45 (1905); R.A.Dyer in Flora S. Afr. 27,4:67 (1980) [Fig.102, 102A]

Perennial dwarf herb with tuber. *Tuber* 30-100 mm broad, 20-70 mm high, with central depression. *Stems* 1-several-many produced annually, 60-120 mm

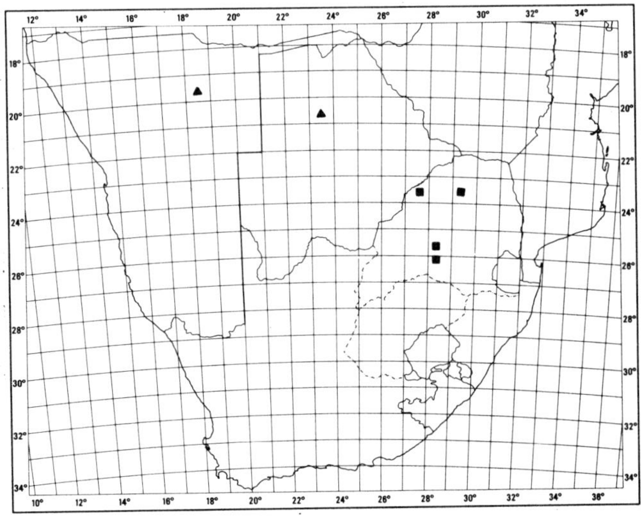

■ *Ceropegia conrathii* (34)
▲ *Ceropegia floribunda* (34a)

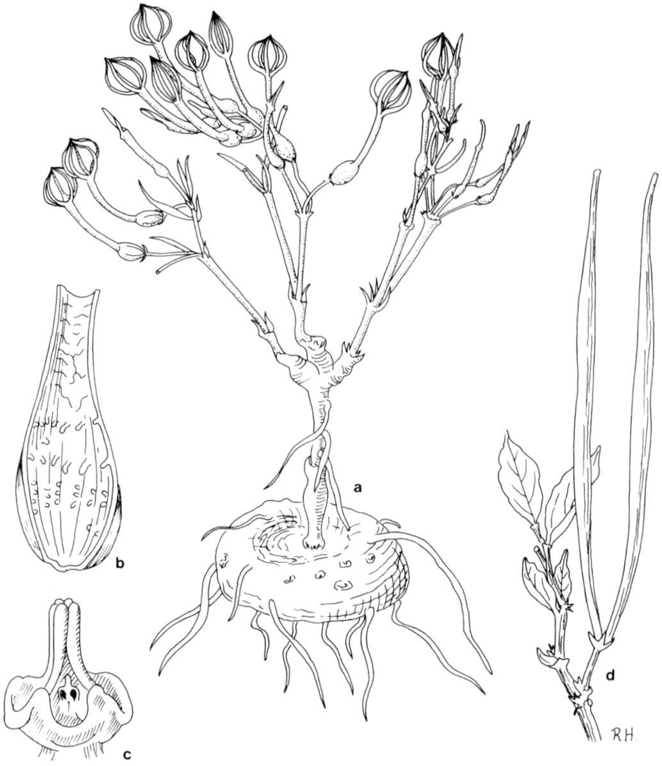

Figure 102. *Ceropegia conrathii:* a, plant, x ½; b, longitudinal section of base of corolla, x 3; c, corona, x 5; d, fruiting branch, x ½ (adapted from Flower. Pl. Afr. T.1246) (Fig.13 in Flora S.Afr. 27,4:(1979)).

tall. *Leaves* up to 30 mm long and 12 mm broad; often with undulate or cripsed margin. *Flowers* sometimes developing before leaves, 1-6 lateral at nodes, often more than one open together. *Corolla* about 35 mm long; tube 20-23 mm long with basal inflation up to 3,5-4 mm diameter, 10 mm long, reduced to about 2 mm diameter and not much expanded at mouth; lobes narrow, about 8 mm long, connate at tips forming cage-like structure; occasionally becoming free, vinaceous cinnamon with a lighter shade on the tube.

Ceropegia conrathii is named after Paul Conrath, chemist, naturalist, who came from eastern Europe as Manager of the dynamite factory at Modderfontein, near Johannesburg early in the 20th century. The type specimen of the species originated near the factory and the species has since been traced at sites near Potchefstroom, Ellisras on the Botswana border, near Pretoria and there is a most perplexing isolated record from near Lala Nek near the coast in Zululand. Mostly the corolla lobes are united at the tips but free lobes do occasionally occur. H.Huber equated this dwarf species with *Ceropegia floribunda,* a robust climber but I do not follow this treatment since no specimen from southern Africa has shown any tendency to produce climbing stems and it sometimes flowers prior to leaf development.

Figure 102A. *Ceropegia conrathii:* a, form very much more branched from base than in the typical form, x $\frac{1}{3}$; b, x 1; c, showing the long slender beak buds and rounded cabe on opening, x $1\frac{2}{3}$ (*L.Fivaz* at Ellisras on Botswana border, cult. BRI, photo. Adela Romanowski).

34a. **Ceropegia floribunda** *N.E.Br.* in Fl.Trop.Afr.4,1:460 (1904); R.A.Dyer in Flora S.Afr. 27,4:81 (1980) [Fig.103]

Perennial herb with tuber. *Stem* annual twining, without hairs. *Leaves* somewhat fleshy with short stalk; blade 15-25 x 8-15 mm, elliptic to elliptic-oblong, obtuse or with recurved sharp apex, minutely and thinly hairy on the undersurface and minutely hairy on the margin. *Flowers* many in dense stalkless clusters lateral at the nodes, each flower on a short stalk. *Corolla* green or yellowish with few hairs within; tube about 12 mm long with globose basal inflation, slender above and not widened above at mouth; lobes united at tips, slender, flat on back.

Figure 103. *Ceropegia floribunda:* vigorous climber with clusters of flowers at the nodes, x ¾ (*R.Story* 6483, northern Botswana, cult. BRI, photo J.Reyburn).

Ceropegia floribunda seems to live up to its name as a free flowering species. It is recorded as rare by the original collector and does not appear to have a wide distribution away from its type locality on the Kwebe Hills in Ngamiland, Botswana. The corolla was described as green with purplish base. With the assistance of David Field at the Royal Botanic Gardens, Kew, I have associated with it Story 6483 which was collected in the Grootfontein district of SWA, about 20 km north of Gautscha Pan in sandy soil with *Acacia* sp., *Lonchocar-*

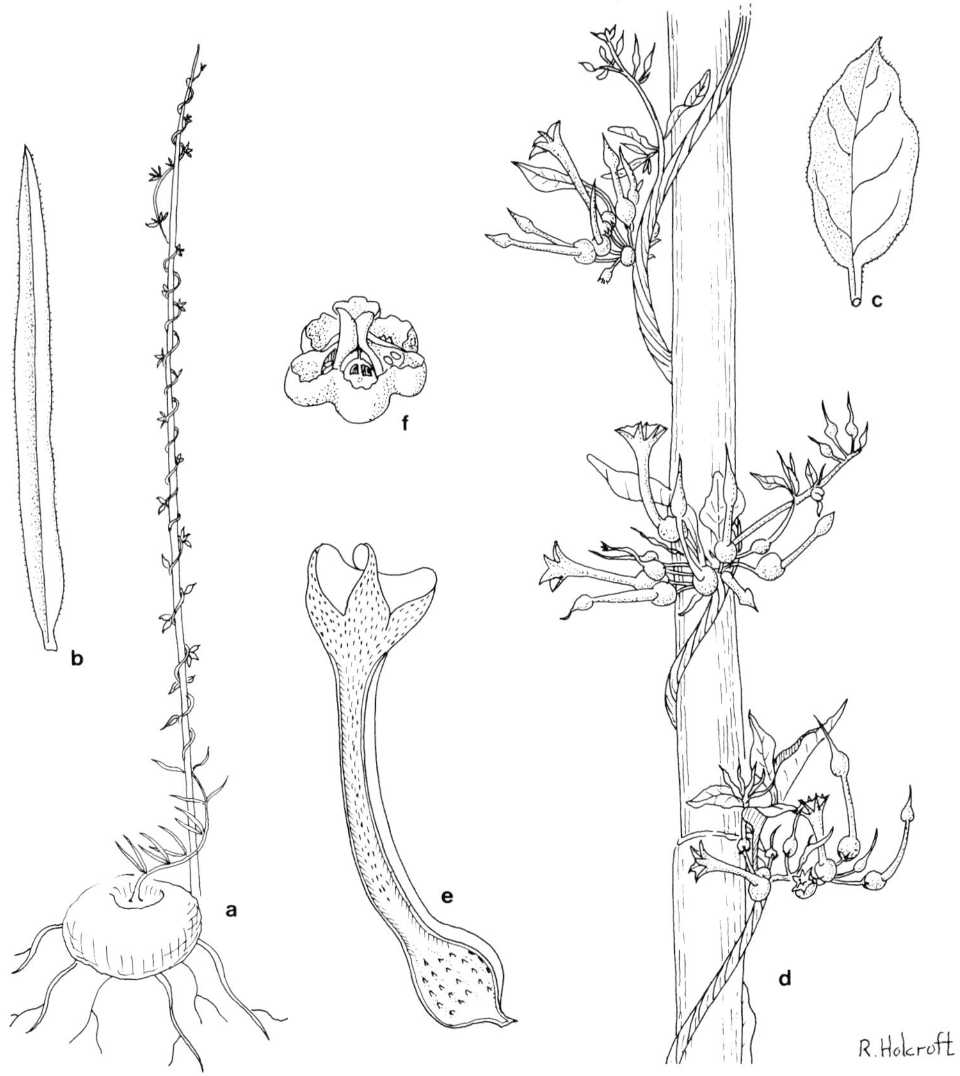

Figure 104. *Ceropegia multiflora* subsp. *multiflora:* a, habit, x ¼; b & c, leaf-shape variation, x 1½; d, twining flowering stem, x ¾; e, longitudinal section of corolla (fused tips of corolla-lobes bent outwards as a result of cutting the corolla-tube), x 3; f, corona, x 7½ (after Flower. Pl. Afr. t.909) (Fig.14 in Flora S.Afr. 27,4:70 (1979)).

pus sp. and *Eragrostis* grass. The tuber of Story 6483 was grown and flowered at BRI.

35. **Ceropegia multiflora** *Bak.* in Refug.Bot.1:t.10 (1969); R.A.Dyer in Flora S.Afr. 27,4:69 (1980) [Fig.104, 104A]

Perennial herb producing a tuber. *Tuber* up to about 120 mm diameter, somewhat depressed and often concave on upper surface. *Stems* produced annually, slender, twining up to about 3 m long. *Leaves* often lower ones slender up to about 70 mm long, upper ones shorter and broader, shortly stalked, fleshy. *Flowers* (1)few-many, clustered at nodes, sometimes more than one opening at a time. *Corolla* in bud with needle-like beak 16-26 mm long, greenish; tube with globose basal inflation 4-5 mm diameter, reduced to 2 mm diameter and only slightly expanded to mouth; lobes becoming very slender.

Ceropegia multiflora is one of the free flowering species and, depending largely on the corolla-lobes being united or free at the apex, two subspecies are recognised.

Corolla-lobes united at tips subsp. **multiflora**
Corolla-lobes free naturally at the tips subsp. **tentaculata**

Figure 104A. *Ceropegia multiflora* subsp. *tentaculata:* filiform tips of corolla lobes free, x 4/5 (*B.Levy* 3822, cult. BRI).

subsp. **multiflora** [Fig.104]
Corolla-lobes from a broad triangular base, abruptly tapering into thread-like tips; tips stretched horizontally over the mouth of the tube, only becoming free with rough handling:

The description of the subspecies was based on a plant grown at the Royal Botanic Gardens, Kew, from a tuber sent by Arnot from Colesberg in the Cape. In Flora Capensis 1908, N.E.Brown suggested that Arnot had collected the type in Griqualand West where he acted as agent to the Griqua Government. A positive record has since been made at Philipstown which is not far remote from Colesberg. The distribution extends from there to SWA and Transvaal.

subsp. **tentaculata** *(N.E.Br.) Huber* in Mems Soc.broteriana 12:122 (1958) [Fig.104A]

Corolla-lobes from a broad triangular base, abruptly tapering into free thread-like tips. The typical form with genuinely free tips was described from Angola. The terminal portion of the corolla-lobes is so thin and fragile that one may be left in doubt whether the lobes are free by accident or design.

36. **Ceropegia paricyma** *N.E.Br.* in Kew Bull. 1898:309 (1898); R.A.Dyer in Flora S.Afr. 27,4:69 (1980) [Fig.105]

Perennial herb with tuber. *Tuber* small. *Stems* annual up to about 1 m tall, twining, sparingly hairy. *Leaves* spreading, stalked up to 20 mm; blade drying thin, often heart-shaped or lobed, about 40 mm long, 30 mm broad. *Flowers*

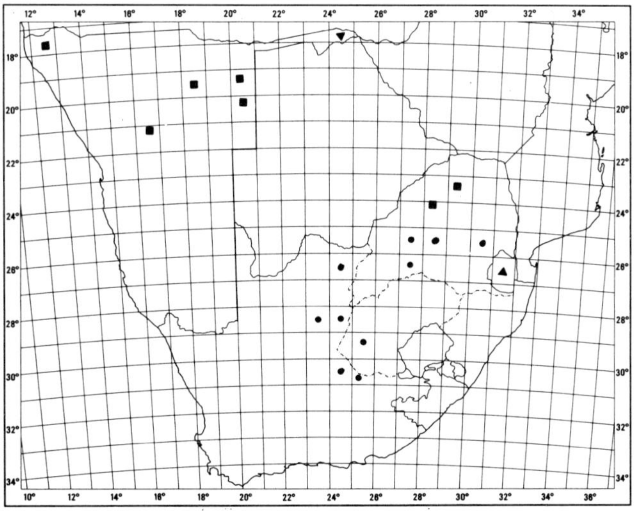

● *Ceropegia multiflora* subsp. *multiflora*
■ *Ceropegia multiflora* subsp. *tentaculata*
▼ *Ceropegia paricyma* (36)
▲ *Ceropegia* sp. (36a)

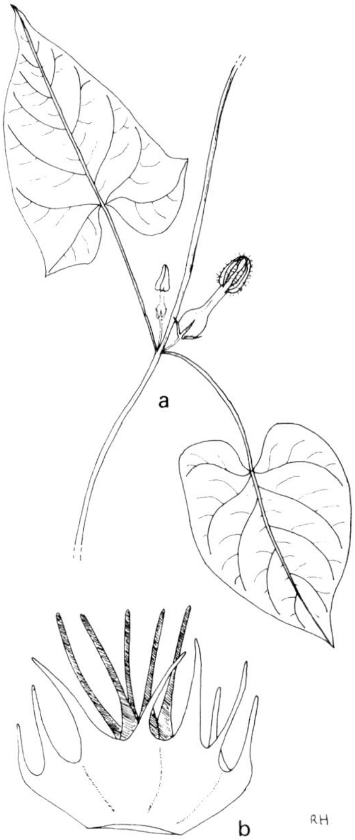

Figure 105. *Ceropegia paricyma:* a, part of twig in flower, x ½; b, corona, x 5 (after *D. V. Field* in letter).

2-6 on same stalk, opening in succession. *Corolla* 16-22 mm long; tube straight, about 15 mm long with globose-inflated base about 5 mm diameter, cream to purple; lobes united at tips, 5-7 mm long, black, strongly recurved margins with long hairs.

Ceropegia paricyma derives its name from the nature of the floral arrangement. It is a tropical African species with one record in the eastern Caprivi collected by *Killick & Leistner, 3349.* It was found to be fairly frequent in bush on rocky ridges.

36a. **Ceropegia** sp.; *R.A.Dyer* in Flora S.Afr. 27,4:82 (1980) [Fig.106]

Perennial herb with tuberous rootstock. *Tuber* more or less globose, about 45 mm diameter, 30 mm high. *Stem* slender, twining, up to 0,7 m long, with internodes 100-190 mm long, hairy. *Leaves* with stalk about 20 mm long; blade lanceolate, about 65 x 20 mm large, heart-shaped at base, thinly hairy on both surfaces, and with minute hairs on margin. *Flowers* usually two open-

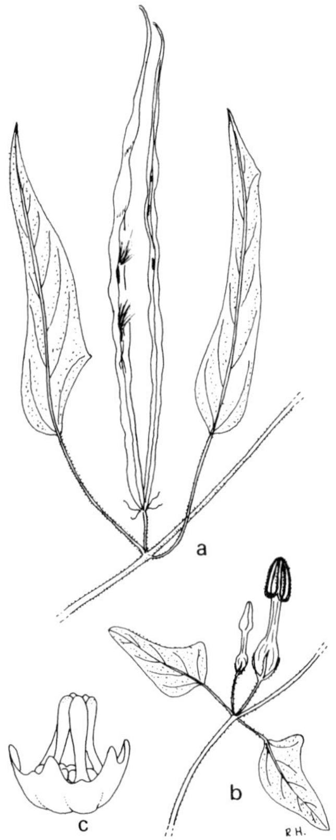

Figure 106. *Ceropegia* sp., a, twig in fruit about to disperse seed, x ½; b, twig in flower, x ½; c, corona, x 5 (after *Kemp* 1300, 662 Swaziland).

ing in succession lateral at the nodes on thin stalks about 10 mm long. *Corolla* 25-30 mm long, globose-inflated at base 6 mm diameter, cylindric above and widening to 2-3 mm at throat, without hairs outside and inside, cream and purple within; lobes united at tips, 6-9 mm long, linear-spathulate, with somewhat reflexed margins, covered with dense purple hairs on inner surface.

This *Ceropegia* has yet to receive a species name and is under investigation by David Field at the Royal Botanic Garden, Kew. It is considered to be closely related to *C.paricyma* N.E.Br. in a group of species not readily defined one from another. Specimens were collected in 1978 by E.S.Kemp in Swaziland 9 km north-east of Mbabane in the Black Umbuluzi River valley.

37. **Ceropegia rendallii** *N.E.Br.* in Kew Bull. 1894:100 (1894); R.A.Dyer in Flora S.Afr. 27,4:71 (1980) [Fig.107]

Perennial herb with tuber. *Tuber* rarely more than 30 mm diameter, sometimes with auxiliary smaller ones. *Stems* very variable in robustness, 1-2 pro-

Figure 107. *Ceropegia rendallii:* very vigorous form from Zoutpansberg, northern Transvaal, showing top of open canopy, x 1½ (*D Hardy* 2713, cult. BRI).

duced annually, on certain exposed rocky outcrops dwarf up to 100 mm non-twining, up to 1 m and more, twining, in subtropical bushveld. *Leaves* somewhat fleshy shortly stalked, the early ones up to 30 mm long and 15 mm broad, soon becoming much smaller on the twining flowering branches. *Flowers* 1-several in succession from short stalks. *Corolla* up to about 25 mm long; mainly greenish; tube up to about 20 mm long, curved or nearly straight, slightly globose-inflated at base, gradually widened to funnel-shaped mouth 5-6 mm diameter; lobes in bud forming a slender cylinder below the unopened canopy, margins folding inwards on opening forming slender struts under the expanded umbrella-like canopy about 8 mm diameter, canopy with 10 deep indentations round the circumference, all margins shortly ciliate with readily detachable hairs.

Ceropegia rendallii is named after Doctor Rendall a resident of Barberton in the eastern Transvaal in 1894 who sent tubers to the Royal Botanic Gardens, Kew. There are some species which will never make the grade in cultivation and there are others which will always remain favourites among collectors. *Ceropegia rendallii* definitely belongs to the latter group because it combines several attractive features. It grows from a small tuber and often produces adventitious tubers.

Its dainty flowers with a small canopy supported by five slender struts never fail to attract attention. There are many growth forms, however, some rather weak growing on the highveld of the Transvaal, but others in more tropical parts are more than adequately vigorous with a long flowering season. The distribution range has been traced from the Transvaal into Swaziland, Zululand, Natal and Maputo. A further record is given as near Parys in the Orange Free State.

38. **Ceropegia barklyi** *Hook.f.* in Curtis's bot.Mag.t.6315 (1877); R.A.Dyer in Flora S.Afr. 27,4:71 (1980) [Plate 18]

Perennial herb with tuber. *Tuber* up to about 50 mm diameter, sometimes producing auxiliary tubers along a rhizome. *Stems* annual, 1-2, often dwarf sometimes elongate and twining up to about 600 mm tall. *Leaves* shortly stalked; blade of lower leaves, 15-40 mm long, 5-20 mm broad, green or mottled above, fleshy. *Flowers* 1-3 on stalk, opening in succession. *Corolla* 25-35 mm long; tube about 15 mm long, with inflated base, 3,5 mm diameter, mouth about 5 mm broad; lobes forming long slender beak in bud, as long or longer than the tube, sometimes slightly twisted and touching in middle, united at tips, forming elongate cage margins recurved, with few long hairs below halfway to mouth.

Ceropegia barklyi is named after Sir Henry Barkly who, when Governor at the Cape in 1877, sent tubers to the Royal Botanic Gardens, Kew, which he in turn had received from Henry Bowker in the Transkei. As in the closely

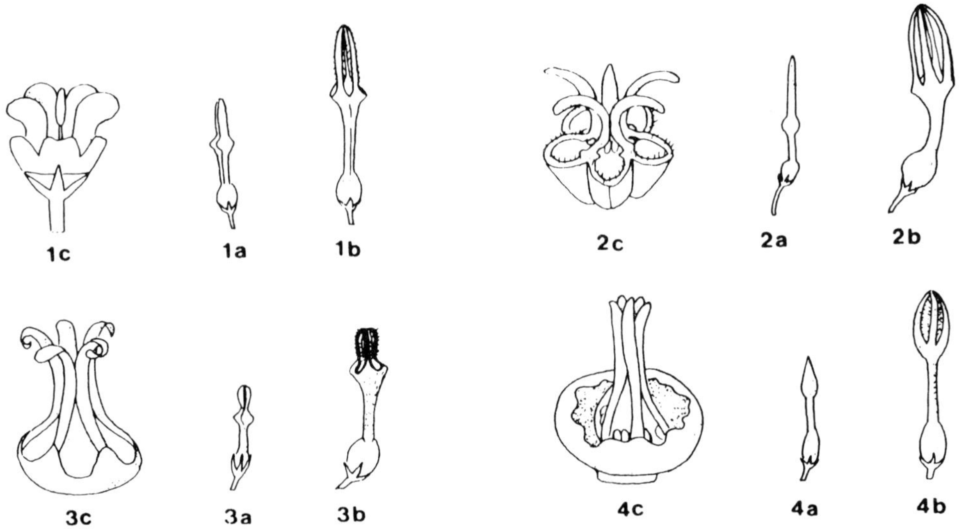

Figure 108. 1. *Ceropegia africana:* 1a, bud, x ¾; 1b, flower, x ¾; 1c, corona with laterally compressed, falcate, obtuse inner lobes, x 6. 2. *C.cancellata:* 2a, bud, x ¾; 2b, flower, x ¾; 2c, corona 5-fluted base with deep pockets and erect, recuring inner lobes, ± triangular in cross-section, x 6. 3. *C.woodii:* 3a, bud, x ¾; 3b, flower, x ¾; 3c, corona with shallowly basin-shaped base and erect-recuring, linear inner lobes, somewhat flattened back and front, x 7½. 4. *C.fortuita:* 4a, bud, x ¾; 4b, flower, x ¾; 4c, basin-shaped outer corona with linear inner lobes arising from within the basin, x 7½ (adapted from PRE records).

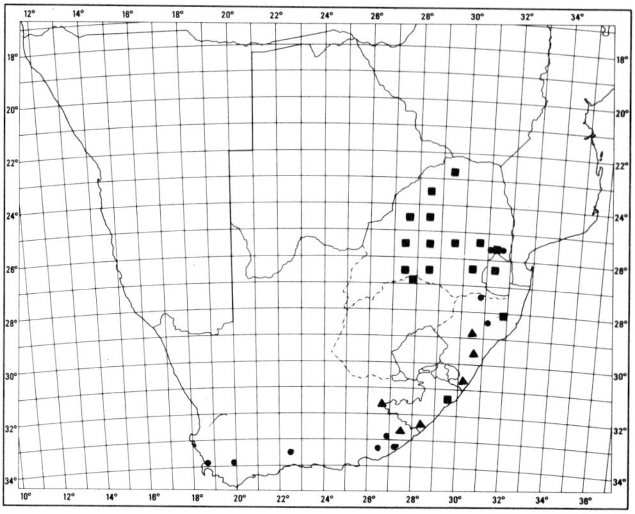

■ *Ceropegia rendallii* (37)
▲ *Ceropegia barklyi* (38)
● *Ceropegia africana* (39)

related species *Ceropegia africana, C.barklyi* occurs in different vegetative forms, from a dwarf spreading, non-twining plant to an elongate twining one up to 1 m or more long. The few long hairs below the middle of the long corolla lobes are of diagnostic value in comparing it with *C.africana* but there does appear to be some overlapping in character. As a general rule the corolla lobes of *C.barklyi* are slender and nearly as long or longer than the tube whereas in *C.africana,* the corolla lobes are broad and markedly shorter than the tube. It is in the short, broad sickle-shape of the inner corona lobes that *C. barklyi, C.africana* and *C.rendallii* are distinguished from other related species.

C.barklyi has been traced in distribution in scrub bush of the eastern Cape from Stutterheim and Queenstown through Natal northwards to the vicinity of Barberton.

39. **Ceropegia africana** *R.Br.* in Bot.Reg.t.626 (1822); R.A.Dyer in Flora S. Afr. 27,4:73 (1980) [Fig.109]

Perennial herb with tuber. *Tuber* about 50 mm diameter, sometimes with auxiliary tubers. *Stems* annual from the tubers, sparsely branched from near base, varying from non-twining dwarf plants to vigorous climbers. *Leaves* somewhat fleshy, shortly stalked, blade up to about 25 mm long and 10 mm broad. *Flowers* 2-several in succession on short stalks lateral at the nodes. *Corolla* 20-25(30) mm long; tube 12-16(18) mm long, straight or slightly curved, with basal globose inflation ±4 mm diameter, reduced to 1-2 mm above

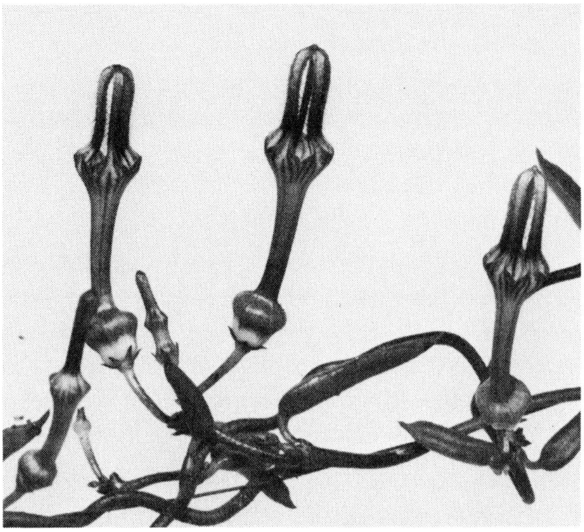

Figure 109. *Ceropegia africana:* typical form, x 1½, near Cango Caves, Oudtshoorn, Cape Province (*E.A.Phillips* 3).

and expanded to about 5 mm at funnel-shaped mouth, greenish striped with violet-brown, lobes linear forming slender beak in bud, connate at tips elongating as bud opens, linear from deltoid base 6-12(15) mm long, margins folded back, connivent-erect from slightly wider spreading mouth, violet-brown on back margins ciliate, inner keels with few dark purple hairs.

Ceropegia africana, as the name suggests, was the first species of *Ceropegia* to be described from Southern Africa. As such it is of special importance to have it accurately circumscribed. The exact locality of the specimen described by Robert Brown in 1822 is not recorded but very similar forms have been collected in the neighbourhood of Oudtshoorn in the southern Cape. The first tubers were introduced to the nursery of Messrs Colvillin, King's Road, Chelsea, London, and illustrated to accompany Brown's description. It is of interest to note that, although described as early as 1822, the species was not recorded from the Cape Peninsula until 1948, when Harry Hall, then a newcomer to South Africa for employment at Kirstenbosch, collected tubers on Lion's Head. Botanists at the Cape were more than sceptical that it occurred there naturally. Harry Hall is too reliable a collector to doubt and the occurrence is admittedly very limited, but distinctive in its dwarf habit. Another distinctive more curious prostrate form has quite recently been recorded from the southern Cape by Bruce Bayer. As recorded in subtropical eastern Transvaal, it is a vigorous climber. Throughout this wide distribution range the laterally compressed sickle-shaped inner corona lobes are remarkably constant in character. This form of inner corona lobe is shared with the closely related *Ceropegia barklyi* (38) and the distinctive *C.rendallii* (37).

40. **Ceropegia occulta** *R.A.Dyer* in Bothalia 7:21 (1958); R.A.Dyer in Flora S.Afr. 27,4:73 (1980) [Fig.110]

Perennial herb with tuber and sometimes with auxiliary tubers. *Tubers* subglobose, 15-40 mm diameter. *Stems* annual, subsucculent, sparingly branched, straggling, very seldom showing signs of twining; up to about 200 mm long. *Leaves* from nearly circular towards base of stems to slender on terminal growth, 10-20 mm long, 3-10 mm broad, fleshy, grooved down face. *Flowers* 1-3 in succession. *Corolla* 20-28 mm long, usually very slightly curved; tube 15-20 mm long, with slightly inflated subglobose base about 5 mm diameter, narrowing to 2,5-3,0 mm and expanding to mouth 5-6 mm diameter, finely longitudinally ribbed within inflation; lobes 5-8 mm long, about 3,5 mm broad at base, contracted to 2,5 mm near middle, united at apex for 2 mm forming

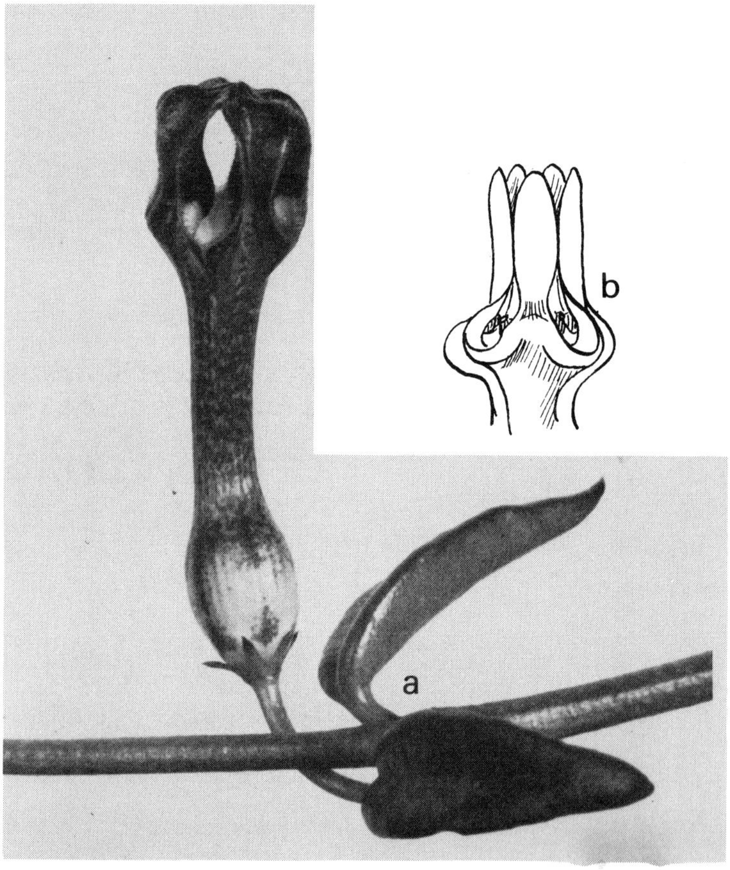

Figure 110. *Ceropegia occulta:* a, flower showing an external similarity to *C.connivens* also from Worcester, Cape, but with tuber as rootstock and no clavate hairs on corolla lobes, x 3; b, corona with outer margin forming 5 pockets and confluent with base of inner lobes, x 10 (*P.A.B.van Breda* 85, type illustration).

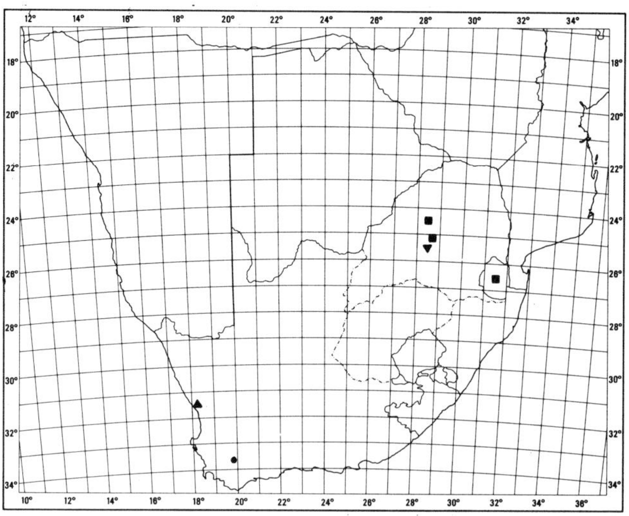

● *Ceropegia occulta* (40)
■ *Ceropegia decidua* subsp. *decidua*
▼ *Ceropegia decidua* subsp. *pretoriensis*
▲ *Ceropegia occidentalis* (42)

small canopy about 4 mm diameter, apiculate, shortly pubescent within the slightly recurved margins.

Ceropegia occulta, which is hidden in its natural environment amongst karroid shrublets, is obviously one of the rare species of the genus. In the 20 years that it has been known it has been collected only a few times and always in the Worcester district of the Cape near Tweefontein.

Without knowledge of the rootstock and corona one might be misled into thinking that this species has a relationship with a form of *Ceropegia connivens* (11) which seems also to be restricted in its natural distribution to the Worcester district of the Cape.

41. **Ceropegia decidua** *E.A.Bruce* in Bothalia 6:213 (1951); R.A.Dyer in Flora S.Afr. 27,4:74 (1980) [Fig.111, 111A]

Perennial herb with tuber. *Tuber* subspherical, somewhat depressed, about 30 mm diameter, sometimes with auxiliary tubers. *Stem* usually single, not or only sparingly branched from near base, a weak climber up to 400 mm tall. *Leaves* shortly petiolate, fleshy, up to 20 mm long, 10 mm broad, sometimes falling before the production of flowers. *Flowers* 2-4 developing in succession. *Corolla* erect, 22-25 mm long; tube 16-19 mm long with subglobose basal inflation 5-7 mm long, 3,5 mm diameter, tuberculate within, slender above, not expanded at mouth, greenish purple, lobes spathulate or subspathulate.

There is some variability within this complex and to provide for it two subspecies are recognised thus:

Figure 111. *Ceropegia decidua* subsp. *decidua:* a, type branch flowering without leaves, lobes slightly spathulate (broadest towards apex), x 1¼; b, rudimentary peduncle, pedicel ± 10 mm, x 5 (*Eric Codd* 47, Warmbaths, north of Pretoria, cult. BRI, photo H.King).

Leaves fairly early deciduous, often before flowering; corolla-lobes united into a small canopy, stems sometimes minutely scabrid — subsp. **decidua**

Leaves persistent for the season; corolla-lobes slightly broadened to the united tips or linear, not forming a canopy — subsp. **pretoriensis**

subsp. **decidua** [Fig.111]

Ceropegia decidua in its typical form sheds its leaves before flowering. It occurs in the northern Transvaal from near Naboomspruit and sporadically eastwards to the warmer parts of Swaziland. Since the species was described in 1951 an appreciable amount of material has been added to the records.

Figure 111A. *Ceropegia decidua* subsp. *pretoriensis:* a, flowering plant with few basal leaves remaining, x 1 (*D.Hardy* 1254, Meintjies Kop, Pretoria, cult. BRI); b, x 1½ (*L.E.Codd* 9565, Meintjies Kop, cult. BRI, photo H.King).

What seemed a constant character in the early collections, namely the falling of the leaves before flowering, later was found to vary in different circumstances. This indicated the relationship with the following subspecies *pretoriensis.*

subsp. **pretoriensis** *R.A.Dyer* in Bothalia 12,4:630 (1979); Flora S.Afr. 27,4: 75 (1980) [Fig.111A]

Ceropegia decidua subsp. *pretoriensis* is found among rocks on hills in the environs of Pretoria. Eileen Bruce had this taxon under observation for several years after 1948 but made no definite decision as regards its classification even when describing the species in 1951. At the present stage there seems no justification for giving this form more than subspecific rank and some workers might consider this unjustifiable.

42. **Ceropegia occidentalis** *R.A.Dyer* in Bothalia 12,3:445 (1978); R.A.Dyer in Flora S.Afr. 27,4:75 (1980)

Perennial herb producing a tuber. *Tuber* 20-40 mm diameter, sometimes with auxiliary ones. *Stem* single or 2 produced annually, 150-300 mm high, with support of shrublets, not twining, somewhat fleshy, sparsely branched, glabrous. *Leaves* fleshy, up to about 20 mm long and 5 mm broad, rounded on back, channelled down face, glabrous, reduced on terminal growth. *Flowers* 2-few opening in succession. *Corolla* 25-30 mm long, with basal swelling 5-7 mm diameter, contracted above and then expanding to mouth about 7 mm wide, pale green with crimson spots and longitudinal markings within; lobes about 7 mm long, united at tips, margins recurved strongly near base; somewhat spathulate above with long dark inflexed hairs.

Ceropegia occidentalis has the most western habitat of all species in Southern Africa. When Harry Hall first sent material of this species from near the mouth of the Olifants River on the West Cape coast, it seemed to be closely related to *Ceropegia africana* but the shape of the corona disproved this. It shows an affinity with *C.occulta,* but the corolla lobes are much narrower. It appears not to have any really close relative, which is not greatly surprising considering how isolated it is in distribution from all other species. Harry Hall did not find it in any other locality along the coast and not very abundant among shrublets at the river mouth.

But to indicate how complicated the questions of habit and habitat are, as I prepare for press, I received the following note from Harry Hall. 'Three years ago I removed a small lateral tuber to give it horticultural treatment, not having seen flowers in the wild for 8-9 years. At Hermanus (C.P.) it grew apace, quickly coiled its stems amongst the supporting twigs and produced flowers which are identical with the Papendorp plants. Further it has flowered for more than 2 years without a pause, having an average of 12 flowers daily. I have bothered to dwell upon this little thing because it shows that some man-made conditions suit some plants much more favourably than nature does'. It emphasises for me how little we know about the inner workings of nature.

43. **Ceropegia cancellata** *Reichb.,* Ic.Bot.Exot.t.207 (1830); R.A.Dyer in Flora S.Afr. 27,4:75 (1980) [Plate 19, Fig.112]

Perennial herb with tuber. *Tuber* becoming rather large, subglobose or oblong, dark brown. *Stem* single or possibly 2 produced annually sometimes branched sparingly, slender, over 300 mm long, climbing-twining. *Leaves* shortly stalked, up to about 25 mm long, 20 mm broad, fleshy with or without hairs. *Flowers* 1-few produced successively. *Corolla* acuminate in bud, 23-35 mm long; tube with globose base, cylindric above, slightly curved, expanded to a trum-

Figure 112. *Ceropegia cancellata:* typical form, a flowering stems showing characteristic long pointed bud and open corolla, x ½; b, leaf, x 1½; c, corona with deeply fluted outer corona lobes forming deep pockets with inner lobes arising more or less from inner rim of tube, x 10 (after *Reichenbach* Ic Bot. Exot. t.207).

pet-like mouth; lobes 10-15 mm long, slightly shorter than tube (typical) united at tips, linear, margins strongly recurved, forming open lattice-work-like cage.

Ceropegia cancellata was given this name meaning latticed referring to the apical cage of the corolla. Unfortunately the precise history of the discovery of the original plant is unrecorded. The first tubers were received in Europe from the Cape before 1830 with the information that they represented an edible Asclepiad. They were grown in the garden of Graf Johan C.von Hoffmannsegg of Dresden. One tuber was sent in flower to his friend Reichenbach for classification. He gave it the name *C.cancellata* but unfortunately his work was overlooked until Bullock drew attention to it in Kew Bulletin 1955:625 (1956). By this time several related species had been described and named and the problem has been to decide on the interrelationships and priorities of these names.

On the question of the place of origin the eastern Cape seems as good a guess as any other and this is supported by the fact that a painting by Harry Bolus of a fresh plant sent to him in 1911 by Florence Paterson of Redhouse in the eastern province near Port Elizabeth, is the nearest likeness seen to the illustration of *C.cancellata.* The Bolus painting is supported by further specimens from F.Paterson in 1917 and 1918. I have gone into further detail on the problems in the Flora of Southern Africa mentioning that H.R.Brownlee when living in Peddie in 1935 made a large number of paintings of the local Ceropegias. Unfortunately he made no dissections and his voucher material was rather poor because of circumstances beyond his control. Some of the illustrations were confirmed as *C.africana* while others were very like *C.cancellata* but there was no definite link-up with a voucher specimen.

Now, at the eleventh hour, Mr and Mrs C.Rostance, friends of Auriol Batten, seem to have confirmed the presence of *C.cancellata* at the Chalumna causeway in the Peddie district which supports the identity of some of Brownlee's paintings.

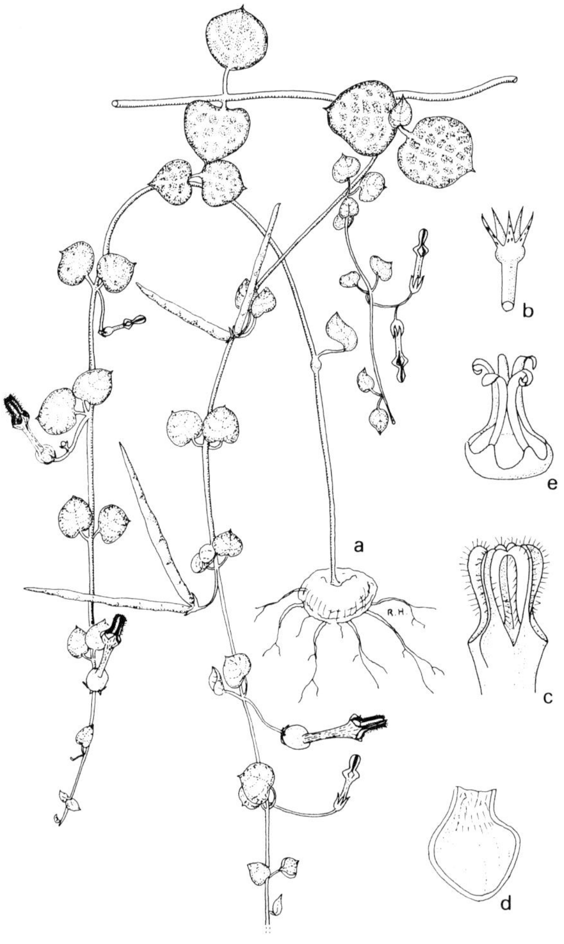

Figure 113. *Ceropegia woodii:* a, plant of typical growth with pendulous perennial stems in fruit and flower, x ½; b, calyx, x 1; c, corolla lobes united at tips, x 2; d, base of corolla tube, x 2; e, corona with shallow outer tube and slender erect-recurved inner lobes, x 5 (after Flower. Pl. Afr. t.382).

Plate 18. *Ceropegia barklyi* (species 38); a. plant with tuber, open flower and bud with corolla lobes twisted, x 1; b. longitudinal section of tubular portion of corolla, x 6; c. corona with fluted base and broadly falcate inner lobes, x 12; d. pollinia, short connectives and wingless carrier (*Auriol Batten,* Butterworth)

b
d
a
c
Auriol Batten

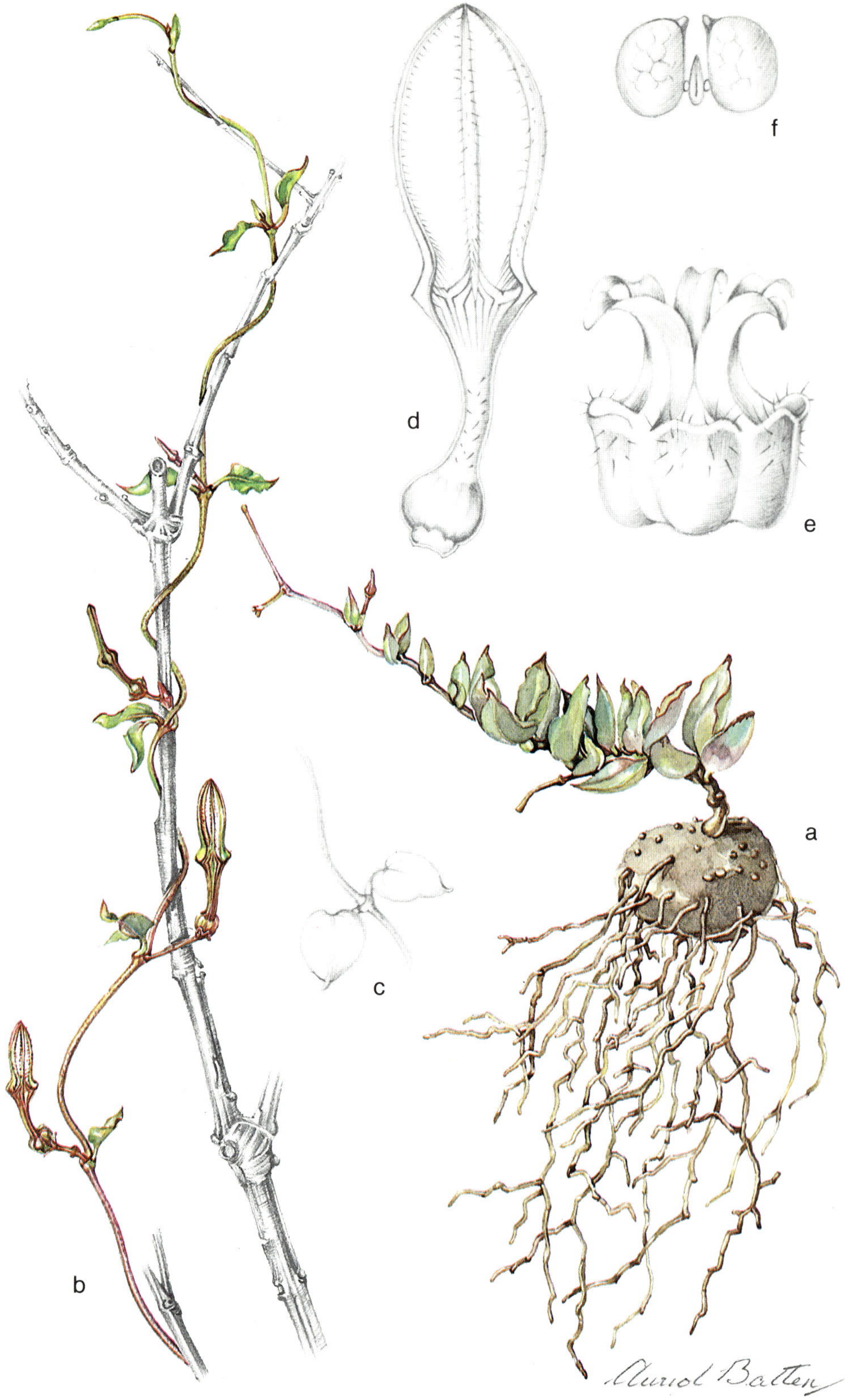
a
b
c
d
e
f
Auriol Batten

44. **Ceropegia woodii** *Schltr.* in Bot.Jb.18, Beibl.45, 34 (1894); R.A.Dyer in Flora S.Afr. 27,4:76 (1980) [Fig.113, 113A]

Perennial herb with tuber. *Tuber* about 25 mm diameter, often with adventitious ones developing at nodes along branches. *Stem* perennial from the base or possibly sometimes annual in the wild, slender, pendent, prostrate, trailing, climbing, glabrous, 25-100 mm long (indefinite in cultivation and under protection). *Leaves* fleshy, shortly stalked; blade 6-18 mm long and broad, uniform green or whitish varigated along the veins above. *Flowers* 1-few developed

Figure 113A. *Ceropegia woodii (= C. euryacme* and *C.barbertonensis):* pendulous flowering stems with upturned flowers, x 1¼ (*E.A.Phillips* 7 from Magoebaskloof, northern Transvaal).

Plate 19. *Ceropegia cancellata* (species 43) a. base of plant with tuber, x 1; b. flowering stem, x 1; c. leaves on upper stem, x 1; d. longitudinal section of corolla (differing in the hairs on the lobes, absent from type), x 4; e. corona with fluted tube and incumbent-erect-spreading inner lobes, x 15; f. pollinia with short connectives to carrier, x 50 (*Auriol Batten,* Chalumna near East London, Cape)

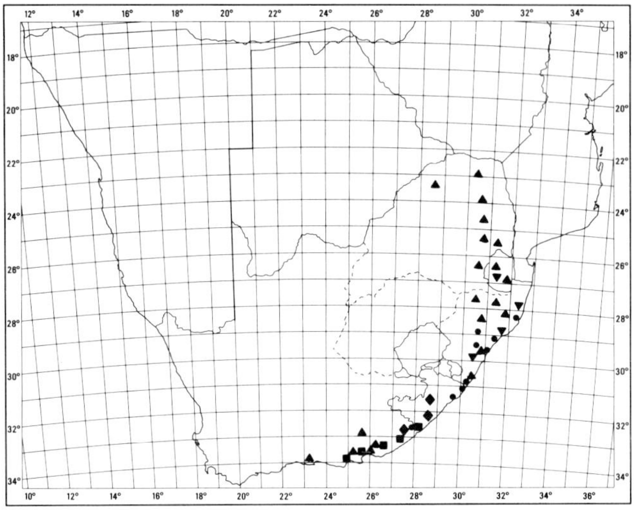

■ *Ceropegia cancellata* (43)
▲ *Ceropegia woodii* (44)
▼ *Ceropegia fortuita* (45)
● *Ceropegia linearis* (46)
◆ *Ceropegia tenuis* (47)

in succession. *Corolla* 18-25 mm long, pinkish on tube with dark striations to the base and the lobes with blackish margins; tube with globose basal inflation 4 mm diameter, contracted to 2 mm and expanded to mouth; lobes 6-7 mm long, narrowly spathulate, with reflexed margins, forming oblong cage with projecting angles at base, with long crinkled purple hairs from inside.

Ceropegia woodii named after Medley Wood founder and director of the Natal Herbarium, Durban. At one time the name *C.woodii* denoted a plant with variegated leaves and stems dangling from a basket. The picture has changed over recent years by the discovery of one variant after another, even to climbing stems, and the consequent sinking into synonymy of several other epithets.

With the present conception of the species the distribution ranges from the mountains of northern Transvaal to the eastern Transvaal into Natal midlands and thence to the Cape Province where it is rather sporadic. Again at the eleventh hour, Bruce Bayer writes that he confirms that P.Bruyns no.1689 from Mossel Bay and also his own specimen no.1566 from Bredasdorp are to be associated with *C.woodii.* Here and elsewhere there is contact with *C.africana* but so far there has been no suggestion of hybridization with this species. There are closer links with *C.linearis* E.Mey. and the synonymy is more fully dealt with in the Flora of S.A. (1980).

45. **Ceropegia fortuita** *R.A.Dyer* in Flower.Pl.S.Afr.24:t.925 (1944); Flora S.Afr. 27,4:77 (1980) [Fig.114]

Perennial herb producing a tuber. *Tuber* up to about 50 mm diameter, slightly

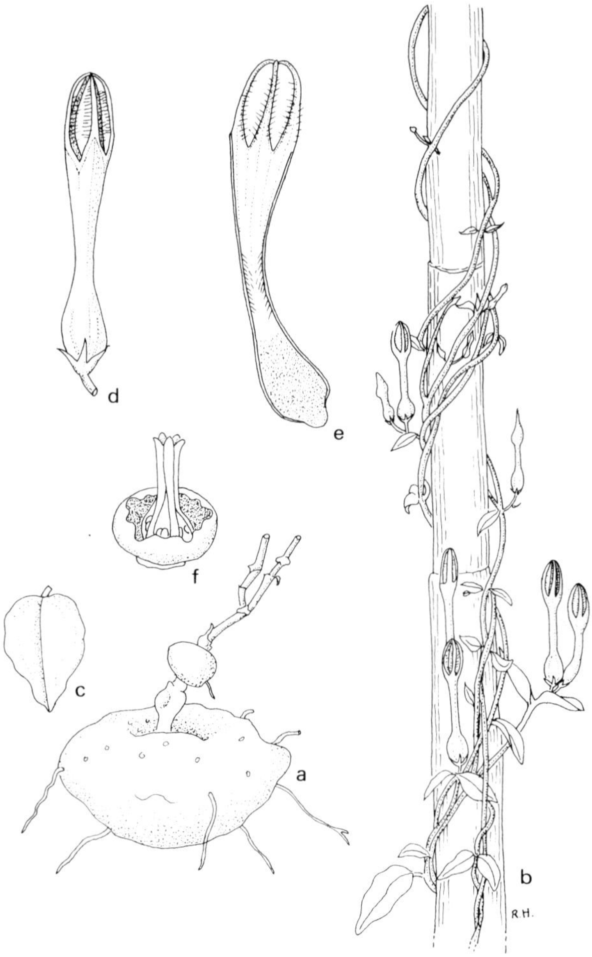

Figure 114. *Ceropegia fortuita:* a, tuber with adventitious tuber above, x ½; b, flowering stems, x ½; c, lower leaf, x ½; d, flower with no broadening of corolla lobes towards the united tips, x 1; e, longitudinal section of corolla, x 1½; f, corona with bowl-shaped outer tube and inner lobes arising from within outer tube, x 5 (after Flower. Pl. Afr. t.925).

depressed, about 30 mm high, sometimes with auxiliary tubers. *Stem* single, annual sparsely branched, dwarf or climbing and twining up to about 1 m tall. *Leaves* shortly stalked, blade up to about 25 mm long and 12-15 mm broad, apiculate, slightly fleshy, usually somewhat deflexed, with few hairs on margin and midrib below. *Flowers* 1-few lateral at nodes, developed successively. *Corolla* in bud with the beak tapering and narrower than the tube, up to about 30 mm long; tubers 18-20 mm long, with basal inflation about 4 mm diameter, contracted to about 1,5 mm and expanded to 2,5 mm at the throat, almost entirely purple within base and purple-striped above, with long hairs and deflexed white hairs within narrow portion; lobes 10-12 mm long united at tips to form narrow cage, minutely ciliate on reflexed margins towards base, with longer dark purple hairs from the keel within.

Ceropegia fortuita was discovered by chance by F.G.Carnegie of Pietermaritzburg, Natal, when collecting succulents in the Valley of 1 000 Hills near Camperdown.

The species is associated with scrubbush in the hot river valleys of Natal and Zululand. It has not been recorded as common nor frequent anywhere. The first tuber was unearthed from between the roots of other plants quite by chance. Although its habit is normally climbing, a dwarf form is on record, which could be most misleading without background information. Two tubers, *Wells* 1365, 1542, were in flower near the type locality during winter at a height of less than 100 mm and no sign of twiners. This is not quite the same as the normally dwarf species *Ceropegia conrathii* which may come into flower in early spring before the growth of normal leaves.

Figure 115. *Ceropegia linearis:* flowering branches of what is regarded as fairly typical for the species, although with broader leaves than type, wide variations do occur in the same geographical area *(A.Mauve-Obermeyer)* Port Edward, Natal, photo H.Lang).

46. **Ceropegia linearis** *E.Mey.*, Comm.Pl.Afr.Austr.194 (1837); R.A.Dyer in Flora S.Afr. 27,4:78 (1980) [Fig.115]

Perennial herb producing a tuber. *Tuber* 20-30 mm diameter, sometimes with auxiliary ones along branches. *Stem* annual or perennial, dwarf to several metres tall or long twining or scrambling. *Leaves* shortly stalked, somewhat fleshy; blade up to 50 mm long and 6 mm broad in typical form but far broader forms are included. *Flowers* 2-3 or more developed in succession. *Corolla* 17-25 mm long; tube with a small globose basal inflation, slender above not expanded at mouth but with slightly spreading sinuses, violaceous striate; lobes narrow, linear or slightly spathulate, margins reflexed with fringe of hairs (pilose) or without hairs.

Ceropegia linearis was so named because of its normally narrow leaves but the interpretation of the species has occupied many man hours of contemplation and field work by several collectors. One would have thought that it would be a simple matter to retrace Drège's footsteps and collect similar if not identical specimens as he did on the coast of Natal south of Port Shepstone over 150 years ago. This has not proved to be the case. For one thing the Drège specimen is incomplete and does not give a clear picture of the plant in its native habitat. Modern collecting by several workers has produced from that general area a wide range of forms which have responded differently under cultivation at BRI, Pretoria. The specimen *Strey* 8077 from that area seems to be the nearest match to the Drège type, but the problem remains, what to include and what to exclude of the many forms and to explain why. One thought is that foreign blood of a related species such as *Ceropegia woodii* may have been introduced at some time, but I have no direct evidence to support this supposition. I have taken a broad view of the species concept because I could find no better alternative.

47. **Ceropegia tenuis** *N.E.Br.* in Fl.Cap.4,1:826 (1908); R.A.Dyer in Flora S. Afr. 27,4:78 (1980) [Plate 20]

Perennial herb with tuber. *Tuber* 15-30 mm diameter, often with smaller adventitious tubers developing. *Stem* slender twining or pendulous, without hairs. *Leaves* shortly stalked; blade somewhat fleshy, up to about 25 mm long, 7-17 mm broad. *Flowers* 3-6, developing in succession. *Corolla* 20-23 mm long; tube 16-19 mm long with small globose basal inflation narrowly cylindric above and only slightly expanded at mouth; lobes about 4 mm long, slightly broadened at united tips, sometimes with few long purple hairs from the keels within.

Ceropegia tenuis is noted for its generally slender habit for it seems fairly constant in character in its distribution area. It was first recorded by Alice

Pegler in the Transkei and has since been traced in the eastern Cape Province in the Komga, King William's Town, East London and Peddie districts associated with scrub growth. Auriol Batten had difficulty in locating a specimen for illustration in the East London district. *C.tenuis* is closely related to *C. linearis* but the shorter corolla lobes and cage appear to be a consistent feature for distinction. It is perplexing, as in some related species, to find stems either twining upwards in scrub or dangling down from rock ledges while the floral structure remains fairly constant.

48. **Ceropegia pachystelma** *Schltr.* in Bot.Jb.20, Beibl.51:47 (1895); R.A.Dyer in Flora S.Afr. 27,4:79 (1980) [Fig.116]

Perennial herb producing a tuber. *Tuber* 40-120 mm diameter depressed, with warty upper surface. *Stem* annual, twining, up to a few metres long, hairy. *Leaves* somewhat fleshy, shortly stalked; blade up to about 50 mm long, 30 mm broad, sometimes undulate, hairy. *Flowers* few-many on flowering stem, developed in succession but sometimes more than one open at a time. *Corolla* straight or somewhat curved, 20-30 mm long, shortly hairy; tube 15-25 mm long, subglobosely inflated at base up to 5 mm diameter, cylindric above, slightly dilated to mouth; lobes connate at tips, 5-10(12) mm long, spreading wider than the mouth, narrow with margins reflexing, often with pinched-in effect near base, with or without hairs outside and inside.

Ceropegia pachystelma has a large fleshy tuberous rootstock from which arise vigorous climbing stems with somewhat succulent leaves. It is readily cultivated and usually generously free flowering. At various times forms of

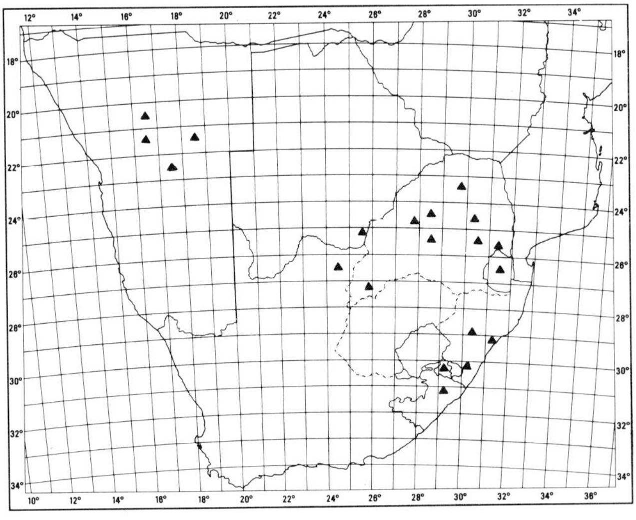

▲ *Ceropegia pachystelma* (48)

this species have had other names including *C.undulata, C.obscura, C.acacietorum* and *C.boerhaaviifolia,* but they have no standing in present circumstances. As understood here, *C.pachystelma* extends from South West Africa to Botswana, northern Cape, Transvaal and Maputo and southwards through Swaziland to southern Natal and Transkei.

Figure 116. *Ceropegia pachystelma:* floriferous plant collected near Umkomaas, Natal, x 2½ (*L.E.Codd* 9369, cult. BRI, photo H.King).

49. **Ceropegia meyeri** *Decne.* in DC.Prod.8:645 (1844); R.A.Dyer in Flora S. Afr. 27,4:79 (1980) [Fig.117]

Perennial herb with tuber. *Tuber* 40-70 mm diameter, depressed, somewhat disc-like. *Stem* annual, herbaceous, twining up to 1,5 m or more, hairy. *Leaves* herbaceous, on stalks 8-30 mm long; blade very variable in outline, 20-50 mm long, 8-30 mm broad, hairy, sometimes with margin entire, sometimes toothed or lobed. *Flowers* 2-4, opening in succession. *Corolla* 40-60 mm long; tube bottle-shaped, with basal two-thirds inflated, to 7-9 mm diameter, narrowed into cylindric neck 2-3 mm diameter, and abruptly dilated to the throat 6-7 mm diameter, with only a few hairs in mouth, white or greenish white on lower half; lobes slender, with reflexed margins, about 10 mm long, connate

Figure 117. *Ceropegia meyeri:* a, freely flowering stem, x ½; b, calyx, x 1½; c, corona, x 5; d, follicle, x ½ (after Flower. Pl. S.Afr. t.30) (Fig.17 in Flora S.Afr. 27,4:(1979)).

for 3 mm at tips, forming a more or less globose cage, velvety black with longitudinal green stripes and long crinkled cilia.

Ceropegia meyeri was given this name by Decaisne in 1844 after E.Meyer, when he found that Meyer had used an illegitimate name for one of Drège's undescribed specimens collected in the Transkei near the 'Basche' river. It has since been traced in its distribution from near Komga northeastwards into Natal, Transvaal, Zimbabwe and Moçambique in association with scrub-bush. There is little cause for misidentifying this species or confusing it with others because of the unique bottle-shape of the inflation, short neck and relatively small roundish cage-like apex with velvety black margins to the struts. The leaves are variable in shape and the whole plant hairy.

50. **Ceropegia stenoloba** *Hochst. ex Chiov.* in Ann.Bot. Roma, 10:395 (1912); R.A.Dyer in Flora S.Afr. 27,4:81 (1980) [Fig.118]

Perrennial herb with tuber. *Tuber* 40-45 mm diameter rounded on top with flattish base. *Stem* annual, slender, twining, 2 or more metres tall, shortly hairy. *Leaves* stalked up to about 70 mm long and 35 mm broad, heart-shaped at base, minutely hairy. *Flowers* 1-many lateral at the nodes, usually more than one open at the same time. *Corolla* about 15 mm long, asymetric; tube slightly curved, about 8 mm long with globose inflated base about 4 mm diameter, cylindric above passing into slightly oblique mouth 2,5 mm diameter, with longitudinal purple markings outside and inside; lobes about 7 mm long with recurved hairy margins united at tips forming subspherical cage curved to one side.

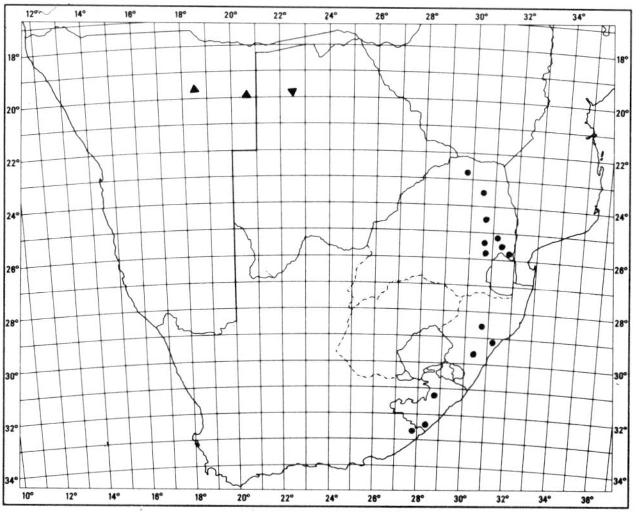

● *Ceropegia meyeri* (49)
▲ *Ceropegia stenoloba* (50)
▼ *Ceropegia bonafouxii* (51)

Figure 118. *Ceropegia stenoloba:* a, flowering plant, x 1; b, flowers exhibiting the unusual asymmetrical structure, x 2 (*R.Story* 5309, Botswana, photo H.King).

Ceropegia stenoloba has narrow corolla-lobes to account for its specific name. The presence of the species in southern Africa is limited to records in northern SWA where it has been located in open woodlands having been traced from Aethiopica southwards. The species is remarkable for the irregularly shaped corolla-cage which is bent to one side of the tube. Many species of *Ceropegia* have a curved or bent corolla-tube but no other species from southern Africa has an asymmetrical cage. The long dark crinkled hairs on the corolla-lobes are also a conspicuous feature of the local form. The distinctive shape of the outer corona-lobes of *C.aberrans* was remarked upon by Schlechter but this is now regarded as equal to *C.stenoloba.*

Figure 119. *Ceropegia bonafouxii,* flowering stems, x 1½ (*De Winter & Marais* 4632, Okavango, S.W.Afr. cult, BRI, photo J.Reyburn).

51. **Ceropegia bonafouxii** *K.Schum.* in Bot.Jb.33:327 (1903); R.A.Dyer in Flora S.Afr. 27,4:81 (1980) [Fig.119]

Perennial herb with tuber. *Tuber* 20-30 mm diameter, sometimes with anchoring roots above tuber. *Stem* annual, slender, herbaceous, twining, 1 m or more tall, moderately hairy with short and longish hairs. *Leaves* herbaceous with longish stalks; blade 30-40(70) mm long, 14-20(25-40) mm broad, hairy more densely so on lower surface. *Flowers* 1-3 lateral at the nodes, opening in succession. *Corolla* bottle-shaped, about 30(40) mm long; tube 22-25(32) mm long, with an elongated basal inflation about 5 mm in diameter, extending more than half its length, contracted into a neck and expanded to the throat about 4 mm diameter, longitudinally purple striate and with purple band round base of neck within; lobes 5(8) mm long, with margins reflexed above middle, united at tips, forming an oblong cage about 7 mm long, velvety-green on inner surface, shortly hairy outside.

Ceropegia bonafouxii was named by Schumann after Bonafoux as a friendly gesture even though the latter was not the collector or discoverer of the type specimen. The flowers are attractive and distinctive among others in southern Africa, but it does not extend far south over the SWA border. Here it is found in sandy soil in *Acacia-Combretum* woodlands, thus it is not readily available unless one has a knowledgeable and willing local contact or obtains material from Angola or Zimbabwe where it is fairly widespread. Sometimes hairiness detracts from the beauty of flowers but in *Ceropegia bonafouxii* this is not the case.

Riocreuxia

Riocreuxia *Decne.* in DC.Prod.8:640 (1844); R.A.Dyer in Flora S.Afr. 27,4: 83 (1980)

Perennial herbs with somewhat woody rootstock, producing a tuft of somewhat fleshy roots. *Stems* mainly annual, twining, suberect and tufted in *R. aberrans*. *Leaves* opposite, stalked, mostly heart-shaped at base and the main ones rather large. *Flowers* usually plentiful, sometimes rather crowded but more usually on branched inflorescences. *Corolla* with usually a subcylindric, slightly inflated tube and lobes united at tips, rarely tube bell-shaped and lobes free at tips (*R.aberrans*).

Riocreuxia is composed of about ten species, one in India and nine in Africa with the main concentration of eight species in Southern Africa. In Southern Africa the distribution is widespread in the Transvaal extending into the northern Cape, OFS, Lesotho, Swaziland, Natal, Transkei, eastern and southern Cape as far west as Mossel Bay. It is noteworthy that there is no record from SWA.

Since the publication of the account in Flora S.Africa (1980), I have reviewed some of my earlier conclusions, mainly as regards the relationships of *R.flanaganii* which was then used as a repository for indecisions. The reassessment is to be found in Bothalia 13,1: (1981) which is followed here. Even so there are several collections in herbaria (mostly BRI) which I have not placed specifically and for which identifications have been left in abeyance.

KEY TO SPECIES OF *RIOCREUXIA*

1 Inner corona-lobes shorter than anthers, sometimes obsolete:
  2 Corolla marked with longitudinal purple, brown or purple-brown veins; bud acuminate-tailed; corona of five broad pockets confluent at base and with ovate outer margin; inner corona-lobes obsolete . . . . 1. **R.picta**
  2 Corolla greenish yellow to white, without longitudinal markings:
    3 Staminal column 1,5-2 mm tall, with five simple, filiform or linear, bifid or toothed, radiating outer corona-lobes (lighthouse-like):
      4 Inner corona-lobes minute or obsolete . . . . . . 2. **R.torulosa**
      4 Inner corona-lobes adpressed to lower half of anther . . 3. **R.bolusii**

3 Staminal column ±1 mm tall; corona with five small pockets with very slightly spreading outer margin, minutely notched with teeth face to face; inner corona-lobes obsolete or to full height of anthers . . . . 4. **R.burchellii**
1 Inner corona-lobes equal to or overtopping the staminal column:
5 Corolla-lobes united at tips; tube subcylindric or somewhat inflated at base:
6 Corolla 16-26 mm long, rarely less; flowers usually many in an inflorescence, 1-3 from nodes of laxly branched cymes . . . . 4. **R.burchellii**
6 Corolla 8-15(16) mm long; inflorescence rarely lax:
7 Outer corona-lobes small, not spreading, 2-toothed; peduncle 0-20 mm long; corolla 10-12(13) mm long; stems bifariously pubescent 6. **R.flanaganii**
7 Outer corona-lobes spreading or spreading-erect; peduncles usually 20 mm or more long; stems usually unifariously pubescent:
8 Outer corona lobes slender, spreading-erect, bifid, equal or overtopping staminal column; corolla 8-10(12) mm long . . . . . 7. **R.alexandrina**
8 Outer corona lobes spreading, subtruncate or bifid, corolla (14)15-18 mm long:
9 Inflorescence of (1)2(3) umbellate clusters of flowers; corolla ±15 mm long; tube with globose base; outer corona-lobes truncate . . . . 5. **R.woodii**
9 Inflorescence usually elongate, several-noded, rarely only (5)20 mm long, few to several flowers from a node; corolla ±17 mm long, with slightly inflated base, outer corona-lobes spreading bifid 4. **R.burchellii**
5 Corolla-lobes free at the tips at maturity; tube campanulate, shorter than lobes, altogether 18-20 mm long, inner corona-lobes considerably overtopping the staminal column 8. **R.aberrans**

SPECIES DESCRIPTIONS AND ILLUSTRATIONS

1. **Riocreuxia picta** *Schltr.* in Bot.Jb.18, Beibl.45:24 (1894); R.A.Dyer in Flora S.Afr. 27,4:84 (1980) [Fig.120]

Perennial climbing herb with rootstock producing a cluster of long fleshy roots. *Stem* produced annually, sparingly branched, twining up to several metres tall over small trees. *Leaves* with stalk up to 50 mm long; blade heart-shaped, sometimes over 150 mm long. *Flowers* many in groups along a branched main axis. *Corolla* 15-25 mm long, with purple, brown, or purple-brown veins; tube somewhat globosely inflated at base, 10-15 mm long; lobes 7-10 mm long, tapering into filiform united tips.

Riocreuxia picta named because of the dark corolla markings extends its distribution from a broad northern Transvaal area southwards to Lesotho, Swaziland, Zululand and northern Natal. It is rather variable in the size of its flowers which may be caused by local habitat conditions, with particular reference to moisture and shade. It may be rare or abundant in different areas.

The identification is made with confidence based on the longitudinal purple, brown, or purple-brown veins in the corolla. The leaves are used by the indigenous population as spinach.

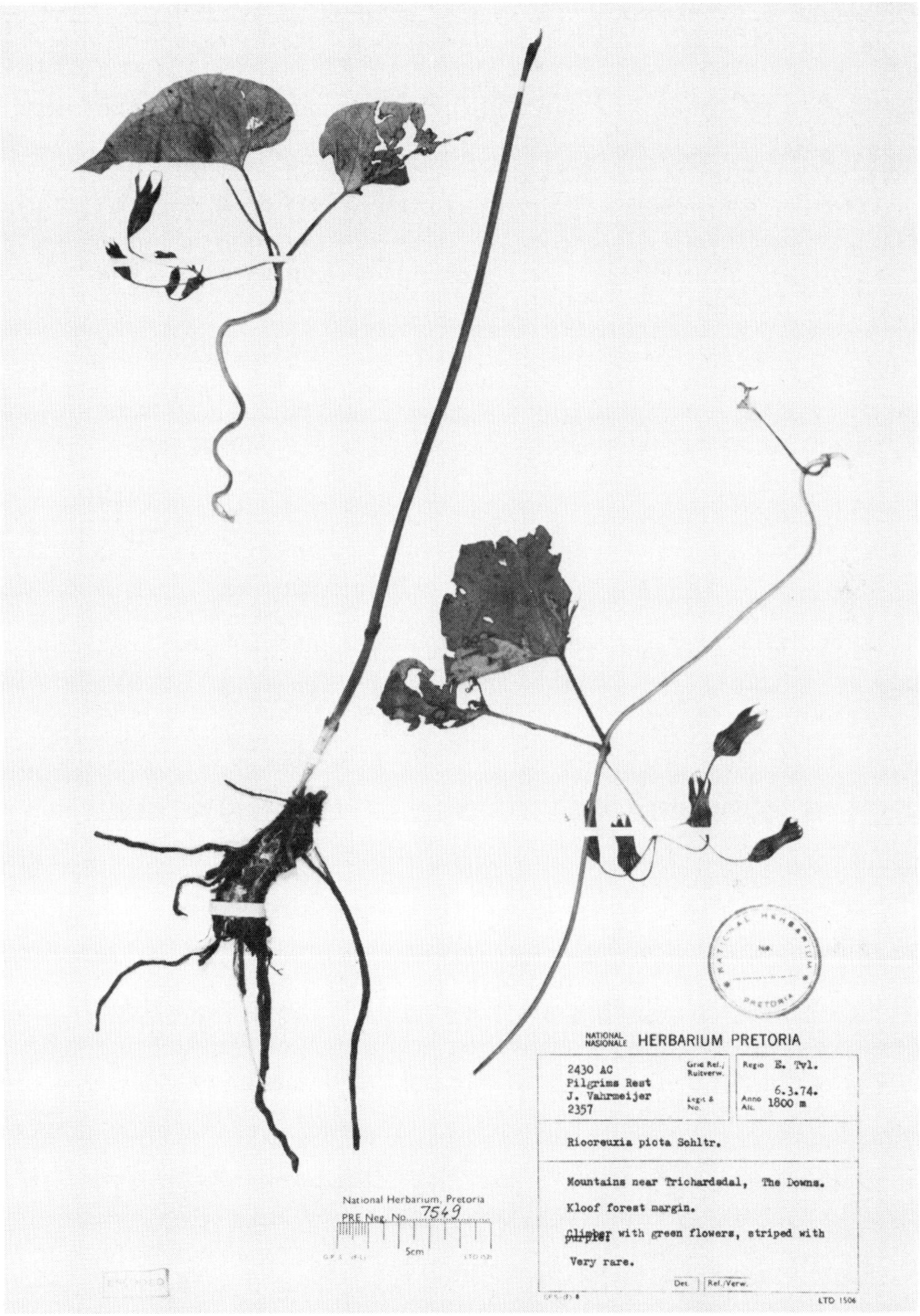

Figure 120. *Riocreuxia picta,* showing the rootstock which is more or less typical for the genus, and rather smaller inflorescences than usual, x ⅓ (*J. Vahrmeijer* 2357, north-eastern Transvaal).

2. **Riocreuxia torulosa** *Decne.* in DC.Prod.8:640 (1844), excl. *Burchell 2668* (= *R.burchellii* K.Schum.); R.A.Dyer in Flora S.Afr. 27,4:84 (1980) [Plate 2]

Perennial herb with rootstock producing a cluster of long somewhat fleshy roots. *Stems* produced annually, freely branched twining up to several metres long. *Leaves* on stalks 15-65 mm long; blade heart-shaped, 30-120 mm long and proportionally broad. *Flowers* in clusters of 2-15, along main stalks up to 100 mm long; individual stalks or pedicels very slender, up to 3 mm long. *Corolla* 12-18(20) mm long, pale yellow tinged with red, green, or purplish; tube somewhat inflated, 6-10(12) mm long; lobes 5-8(10) mm long, slender, tapering to united tips.

Riocreuxia torulosa with its 'bulged' corolla tube is found mainly on forest margins, widespread from northern Transvaal into OFS, Lesotho, Swaziland, Zululand, Natal, into the southern Cape as far west as Mossel Bay.

There is great variability in coronal structure and hairiness within this species concept and N.E.Brown in Flora Capensis 4,1:803 (1908), dealt with this in three varieties. Huber, who included *Riocreuxia* in his revision of *Ceropegia* in 1958, did not uphold Brown's varieties and it seems improbable that a clear-cut subdivision of the herbarium material can be found. Although the corona varies considerably all forms have a characteristic rather long stalk which is usually visible within the light-coloured corolla and has an outline reminiscent of a tall lighthouse.

3. **Riocreuxia bolusii** *N.E.Br.* in Fl.Cap.4,1:802 (1908); R.A.Dyer in Flora S. Afr. 27,4:85 (1980) [Fig.121]

Perennial herb with rootstock producing a cluster of somewhat fleshy roots. *Stems* produced annually, twining, hairy all round. *Leaves* with stalks 10-35 mm long; blade broadly heart-shaped, 35-85 mm long and broad in proportion. *Flowers* numerous in branched inflorescences. *Corolla* apparently yellowish, slightly inflated at base; tube 8-13 mm long; lobes connivent-erect, 6-8 mm long, connate at the tips.

Riocreuxia bolusii is named after Harry Bolus, founder of the Bolus Herbarium, University of Cape Town. The species was based on a single plant which Harry Bolus found on a bank of the Umtata River, near Umtata in the Transkei. In the same environment are also present the two widely spread species *R.torulosa* and *R.burchellii* and the impression is gained that *R.bolusii* may well have arisen by natural hybridization.

Plate 20. *Ceropegia tenuis* (species 47); a. tuber, with lower portion of stem and adventitious tuber, x 1; b and c. middle and terminal flowering stems, x 1; d. adventitious tuber, x 1; e. longitudinal section of basal inflation of corolla, x 4; f. corona with fluted basal tube and inner lobes dorsally flattened, incumbent, erect spreading, x 10; g. pollinia with very short connective caudicles to carrier, x 50 (*Auriol Batten,* 309, near East London, Cape)

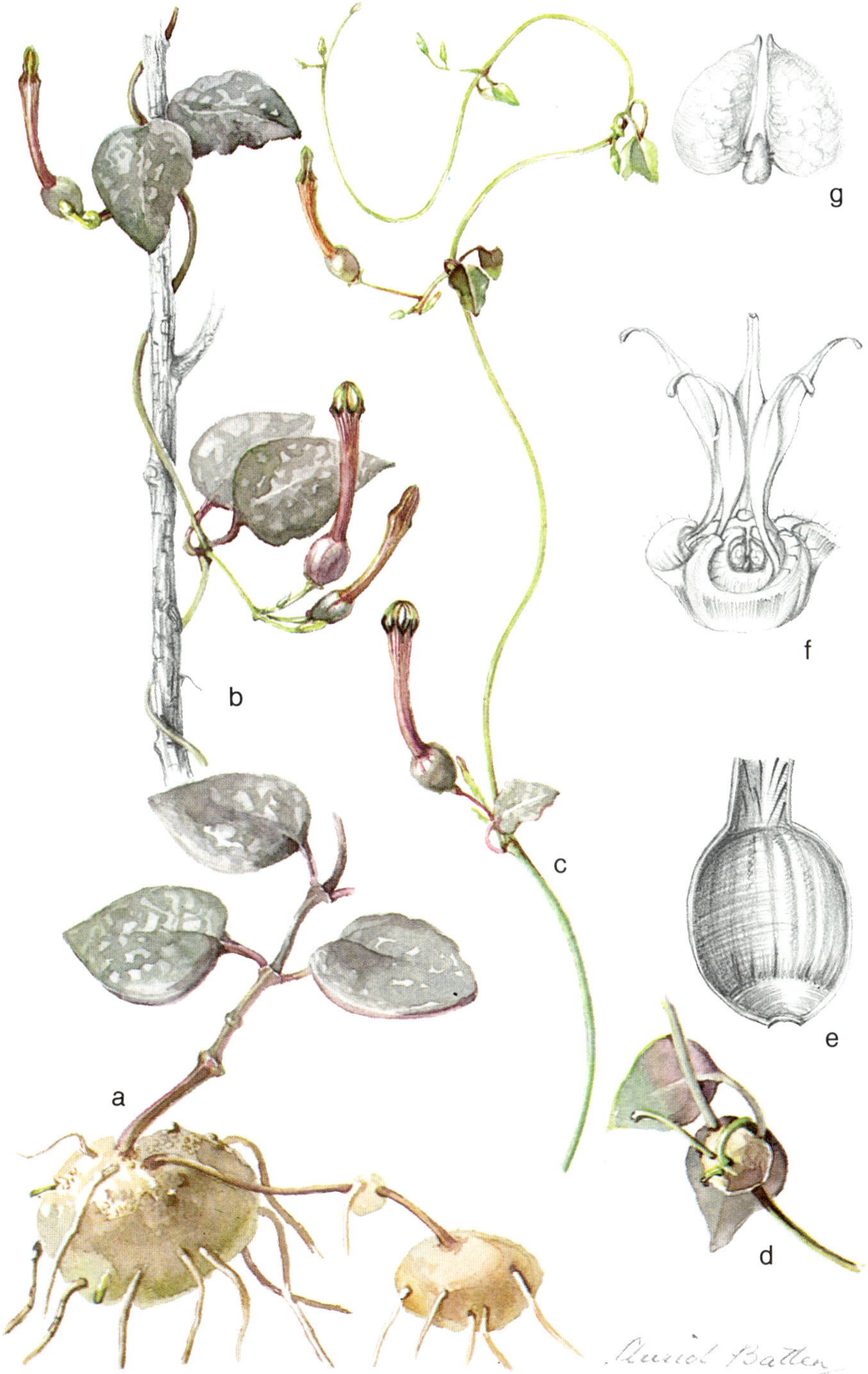
a
b
c
d
e
f
g
Auriel Batten

f
e
c
d
b
a
Auriol Batten

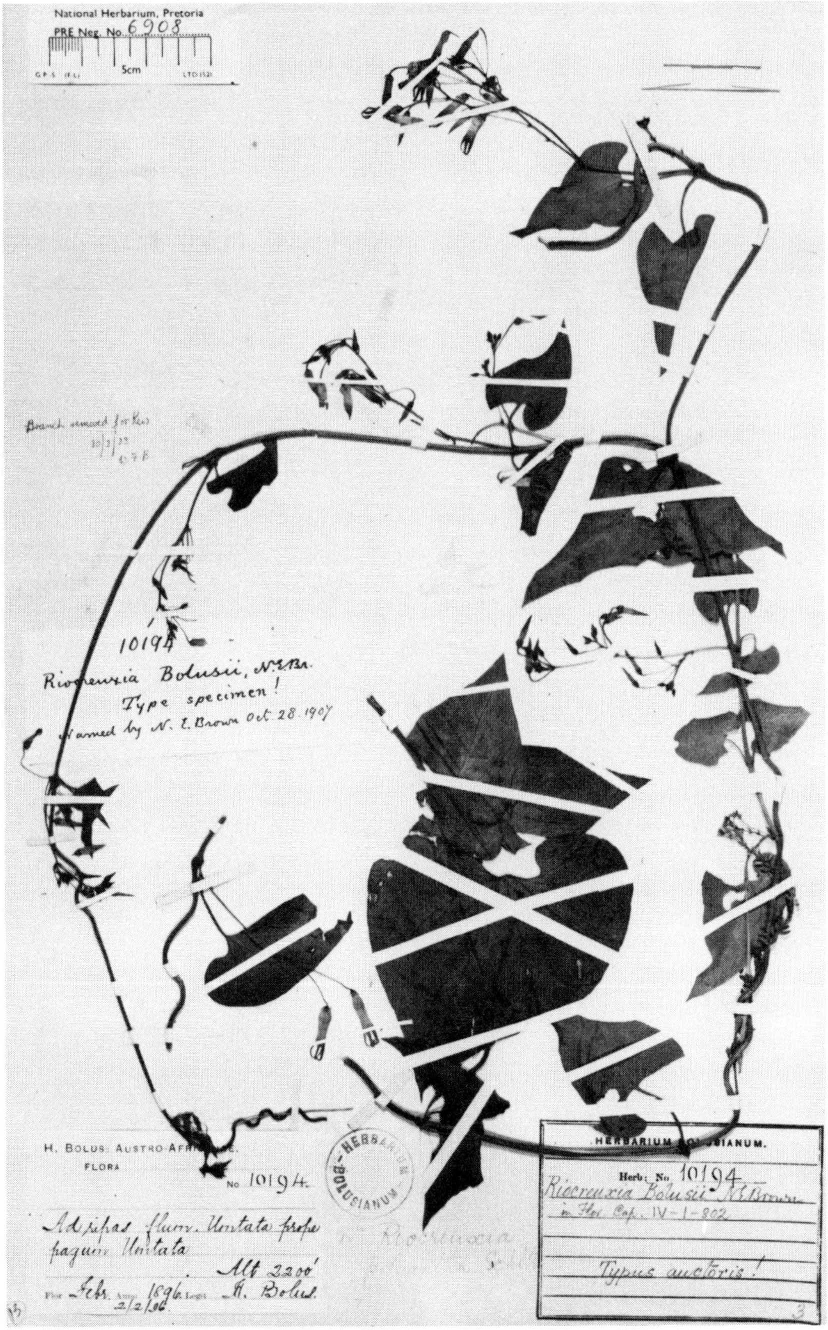

Figure 121. *Riocreuxia bolusii,* photograph of type specimen in Bolus Herbarium, Cape Town.

Plate 21. *Riocreuxia flanaganii* (species 6); a. rootsystem and base of stem, x 1; b. flowering branch, x 1; c. piece of stem with band of hairs between nodes, x 1; d. longitudinal section of corolla, x 5; e. corona with inner lobes overtopping stamunal column, x 30; f. pollinia attached to carrier, x 60 (*Auriol Batten,* West Bank, East London)

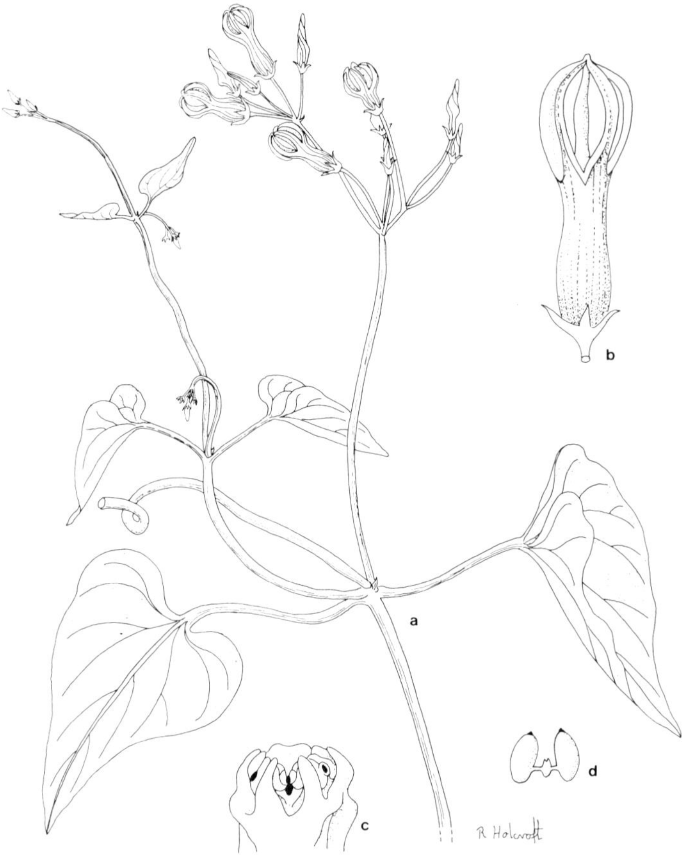

Figure 122. *Riocreuxia burchellii:* a, flowering twig, x ½; b, flower with corolla-lobes united at apex, x 1½; c, corona with 5 pockets and slightly prominent outer margins, inner lobes about equal to staminal column, x 10; d, pollinia with carrier, x 16 (after Flower. Pl. Afr. t.1124).

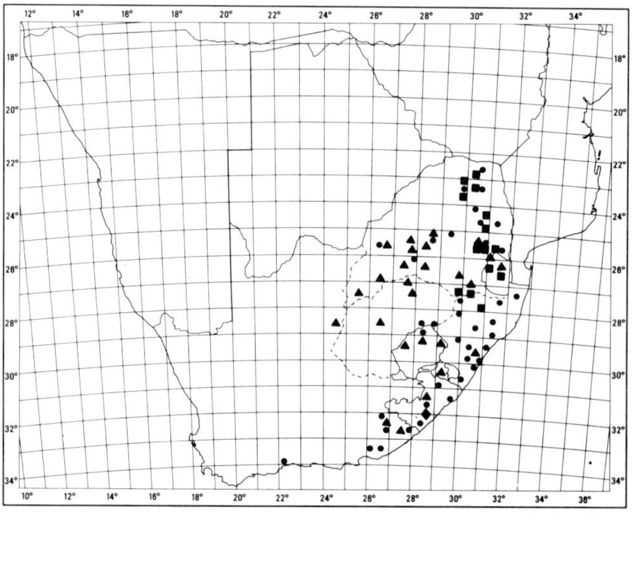

■ *Riocreuxia picta* (1)
● *Riocreuxia torulosa* (2)
+ *Riocreuxia bolusii* (3)
▲ *Riocreuxia burchellii* (4)

4. **Riocreuxia burchellii** *K.Schum.* in Natürl.PflFam. 4,2:273 (1895); R.A.Dyer in Flora S.Afr. 27,4:85 (1980) [Fig.122]

Riocreuxia flanaganii Schltr. subsp. *segregata* R.A.Dyer in Bothalia 12,4:632 (1979), in Flora S.Afr. 27,4:87 (1980).

Perennial climbing herb with rootstock producing a cluster of elongate somewhat fleshy roots. *Stem* produced annually, twining up to several metres long with a single strip of hairs down its length. *Leaves* with stalk 20-50 mm long; blade nearly hairless to thinly hairy, more or less heart-shaped, 30-120(150) mm long, 20-80(120) mm broad. *Flowers* usually many, from long slender peduncles or stalks, which give rise to slender branches bearing the individual flowers. *Corolla* (15)18-26 mm long; tube subcylindric, 10-15 mm long, slightly inflated towards base; lobes 8-12 mm long, tapering to the connate tips.

Riocreuxia burchellii is named after the famous William Burchell, naturalist-explorer in the Cape, 1810-14. It is another widespread species in the northern Transvaal extending into northern Cape, OFS, Lesotho, Swaziland, Zululand and becoming rarer in Natal and eastern Cape. Not infrequently there is an overlapping of the distribution areas with *R.torulosa.* Hybridization does not seem to be a feature of this overlapping but it may have taken place in the production of *R.bolusii. R.torulosa* and *R.burchellii* are readily distinguished by the structure of the corona and *R.torulosa* may be far more hairy.

5. **Riocreuxia woodii** *N.E.Br.* in Fl.Cap.4,1:803 (1908); R.A.Dyer in Bothalia 13 (1981) [Fig.123]

Perennial herb with cluster of roots. *Stems* annual twining with a single row of hairs. *Leaves* with stalks up to about 25 mm broad, thinly and shortly hairy on both surfaces. *Flowers* produced from the nodes on stalks or peduncles 25-80 mm long, bearing (1)2(3) clusters of flowers; clusters 8-24-flowered, somewhat flat-topped. *Corolla* 14-16 mm long; tube 7-8 mm long, slightly globose-inflated at the base, whitish to pale yellow and lobes united at brown tips.

Riocreuxia woodii named after Medley Wood founder of the Natal Herbarium who is mentioned under other species in this publication. The species is recorded by only one collection made in 1879 at Inanda, Natal, some miles inland from Durban. At this locality there is an overlap in distribution with *Riocreuxia torulosa* and it requires careful study of the coronal structure to be sure of the identification. The structure of the inner corona of *R.woodii* shows a closer relationship with *R.flanaganii* but this latter species is many miles to the south-east in the eastern Cape. The habit of the inflorescence is also of importance as a distinguishing character of *R.woodii.* In Flora S.Afr.

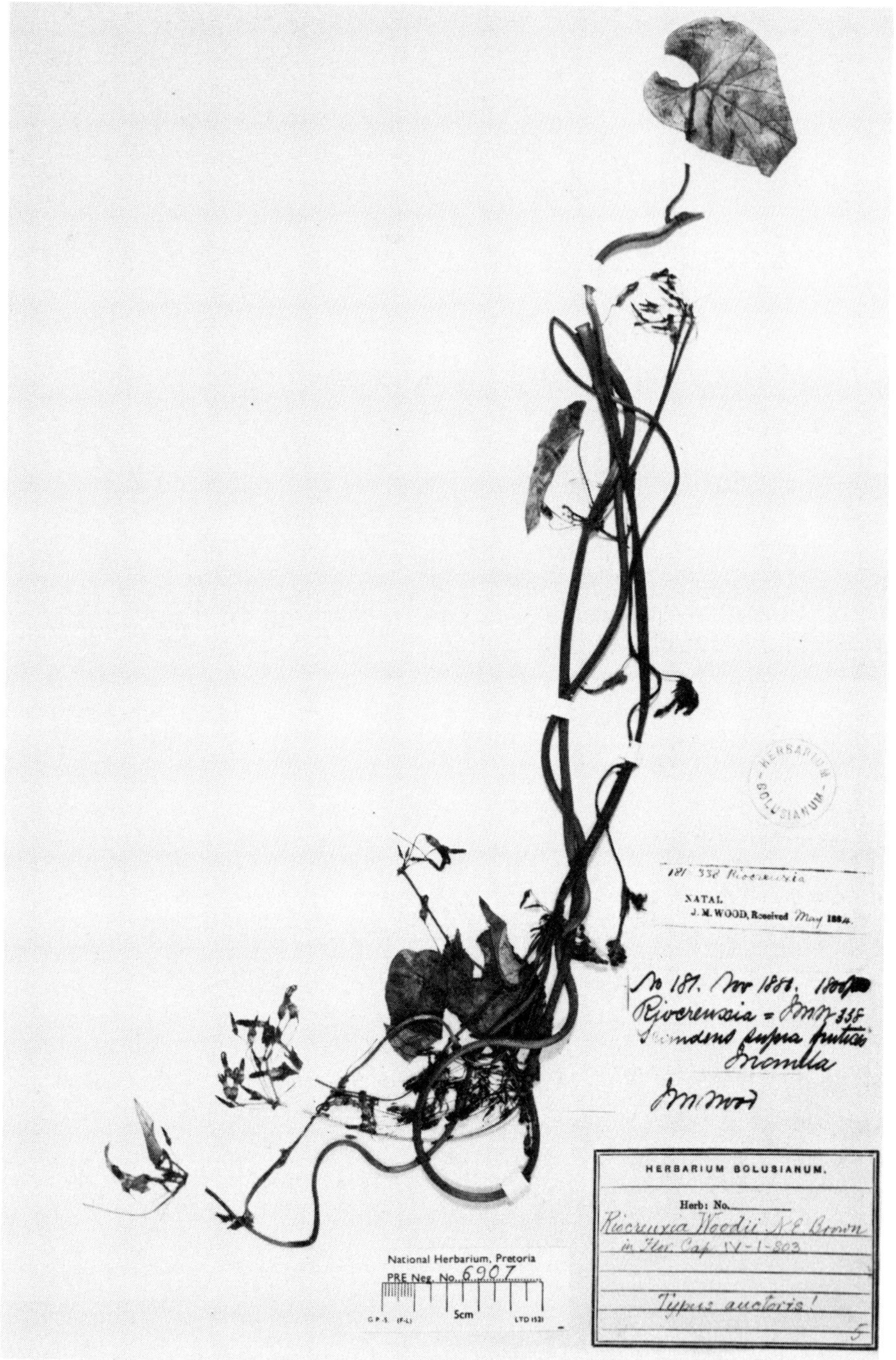

Figure 123. *Riocreuxia woodii,* photograph of the isotype specimen in the Bolus Herbarium, Cape Town.

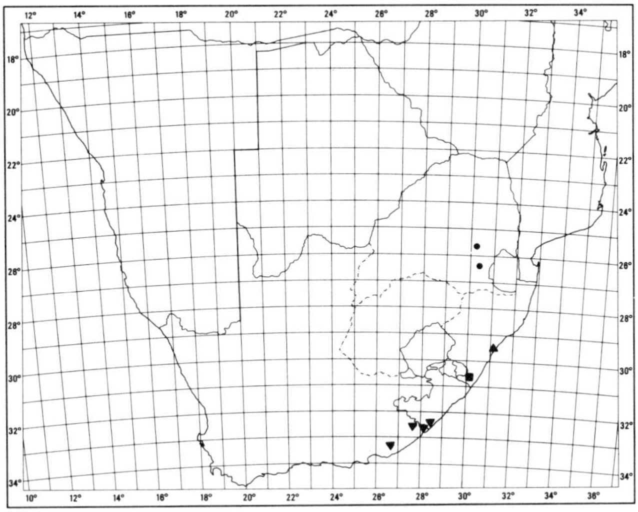

▲ *Riocreuxia woodii* (5)
▼ *Riocreuxia flanaganii* (6)
■ *Riocreuxia alexandrina* (7)
● *Riocreuxia aberrans* (8)

27,4 (1980), *R.woodii* was included as a subspecies of *R.flanaganii.* As indicated above I now consider this to have been an error of judgment.

6. **Riocreuxia flanaganii** *Schltr.*, in Bot.Jb.18 Beibl.45:13 (1894); R.A.Dyer in Bothalia 13 (1981) [Plate 21]

Perennial herb with rootstock producing a cluster of subfleshy roots. *Stems* annual twining up to 1 m or more tall, bifariously puberulous. *Leaves* with stalks up to about 40 mm long; blade more or less heart-shaped, up to about 80 mm long, thinly hairy on both surfaces or nearly hairless. *Flowers* clustered at the stem nodes or clustered on short stalks or peduncles. *Corolla* 10-12 mm long, divided about halfway, pale yellowish on tube and base of lobes, purple-brown above.

Riocreuxia flanaganii named after H.G.Flanagan one of the earliest citrus farmers in the eastern Cape with a very keen interest in Botany. On the present concept of the species, as opposed to that published in the Flora S.Afr. 1980, it is found in the eastern Cape not uncommonly in grassveld and low scrub from south of Grahamstown to East London and Komga and in the southern coastal area of Transkei.

7. **Riocreuxia alexandrina** *R.A.Dyer* in Bothalia 13 (1981). *R.flanaganii* subsp. *alexandrina* (Huber) R.A.Dyer in Flora S.Afr. 27,4:88 (1980)

Perennial herb with a cluster of subfleshy roots. *Stems* twining, with scattered hairs or hairs in single row. *Leaves* broadly heart-shaped at base and narrowed to apex, up to about 75 mm long and 50 mm broad, more or less hairy. *Flowering stalks* short, or slender, up to about 35 mm with rather dense clusters of flowers towards the ends. *Corolla* 8-10(12) mm long, divided half to two-thirds to base; tube slightly inflated, 3-5 mm long; lobes slender united at tips, up to 6 mm long.

Riocreuxia alexandrina is named after the area of southern Natal, inland of Port Shepstone in which Moyeni is included at about 700 m. Plant specimens were collected there by H.Rudatis from 1911 onwards at least some of the *R. alexandrina* being found twining on tall grasses. At one time the Rudatis specimens were regarded as falling within the species concept of *R.flanaganii,* but this was discounted in Bothalia cited above and they now have specific status.

Figure 124. *Riocreuxia aberrans* about 13 km north of Belfast, eastern Transvaal, among rocks in grassveld (photo L.E.Codd).

Figure 125. *Riocreuxia aberrans:* a, habit; b, flowering branch, x ½; c, longitudinal section of flower exposing corona; outer corona forming margin to pockets; inner lobes overtopping the staminal column, x 3½; d, pollinia with shortly pellucid terminal beak, attached basally to carrier, x 20 (after Flower, Pl. Afr. t.1253).

8. **Riocreuxia aberrans** *R.A.Dyer* in Bothalia 3:274 (1937); in Flower.Pl.Afr. t.1253 (1957); Flora S.Afr. 27,4:88 (1980) [Fig.124, 125]

Perennial herb with woody rootstock producing tuft of long, subsucculent roots. *Stems* tufted, erect or somewhat twining, up to 0,6 m or more high, sparsely branched, with internodes 40-50 mm long, unifariously pubescent and with long hairs at nodes. *Leaves* with stalks or petioles 20-40 mm long, heart-shaped up to 50(70) mm long nearly as broad, narrowed to apex, sometimes with scattered hairs and ciliate on margin. *Flowers* many on slender stalks. *Corolla* 18-20 mm long, without hairs, divided far more than halfway; tube campanulate; lobes slender, tips becoming free at maturity.

Riocreuxia aberrans differs from all other species in that the tips of the corolla-lobes become free and spreading on maturity. It has a restricted distribution in the eastern Transvaal on exposed quartzite ridges in the Ermelo and Belfast districts. It is quite attractive when in flower in November and December.

Epilogue

Though I have added many observations not previously recorded, to me personally this account of the three related genera is inadequate, for the simple reason that I am unable to comment with exactitude on the peculiarities of nature. Even had the facilities been at my disposal, I did not have the capacity to undertake the project on a world basis. For those workers who follow, it may be helpful to indicate some of the main trouble spots in our area. A dread of mine is that some over-zealous worker will take it upon himself to resuscitate generic names for distinctive groups of species now united under the umbrella of *Brachystelma.* If this were done it would open the door for the creation of monotypic genera to accommodate unusual species, of which there are many, and to what end?

There are many real problems. One has only to study the distribution maps of related groups such as for *Ceropegia cancellata, C. woodii, C. fortuita, C. linearis* and *C. tenuis* to realize what the chances must be for natural hybridization. For one thing, while the typical forms of each are fairly readily distinguishable, there are many specimens in herbaria which leave one in doubt as to their ancestry.

In attempting to rediscover the typical form of *Ceropegia linearis,* known to have originated from the coastal area of Natal, south of Port Shepstone, I had the co-operation of several workers including R. Strey, H. B. Nicholson and Carole de Wet, the first formerly and the last at present on the staff of BRI. *Strey* 8077 and *Nicholson* PRE 57783 satisfied me that they equalled the type of *C. linearis* near enough to make a positive identification. *De Wet* 168, 193, 194, 195, 196, however, from the rocky slopes above Uvongo beach showed a range of growth and corolla form which indicated a close relationship with *C. linearis* (193, 194) and some with characters of *C. woodii* (195, 196). *C. woodii* is fairly frequent in Natal but the typical form has not been identified from the Port Shepstone coast. There is no way by which I can produce conclusions based on scientific evidence but I feel it incumbent on me to attempt some explanation. The descriptions and keys in the foregoing text give the salient features of affinity and divergence of *C. linearis* and *C. woodii* and the De Wet material seems to combine the two species in different proportions. The variability is such that one does not suggest a simple cross of typical forms but that hybridization took place a long time back and that the present variable population along the south coast of southern Natal in the Port Shepstone district is the result of subsequent segregations and backcrossing and there is very little prospect of stability in present circumstances in this environment or in other similar cases.

I have been criticised for abandoning the subdivision of *Ceropegia* into sections and subsections and for the oversimplification of results to the specific level. My 'Keys' are largely based on what I regard as natural relationships, but relationships are multidirectional in nature and cannot adequately be arranged in a linear series.

Opinions must differ and therein lie the seeds of criticism when desired to germinate them.

When I decided to include this unusual Epilogue, I little thought that I should have to use it as an avenue of material bad news.

While at the lithographers awaiting final copying, seven of the coloured illustrations by Auriol Batten were destroyed in a devastating fire. With great magnanimity and fortitude, the artist undertook the arduous task of replacing in record time the destroyed plates by the use as guides of her original sketches, notes and colour slides: this in spite of a tight schedule of personal commitments including the celebration of her mother's 96th birthday. My gratitude is unbounded and the reader likewise will admire and appreciate such an effort which has minimised the consequent inevitable delay in publication.

Incidentally, the publication of the work suffered an initial setback when I had a stroke in 1981.

Index

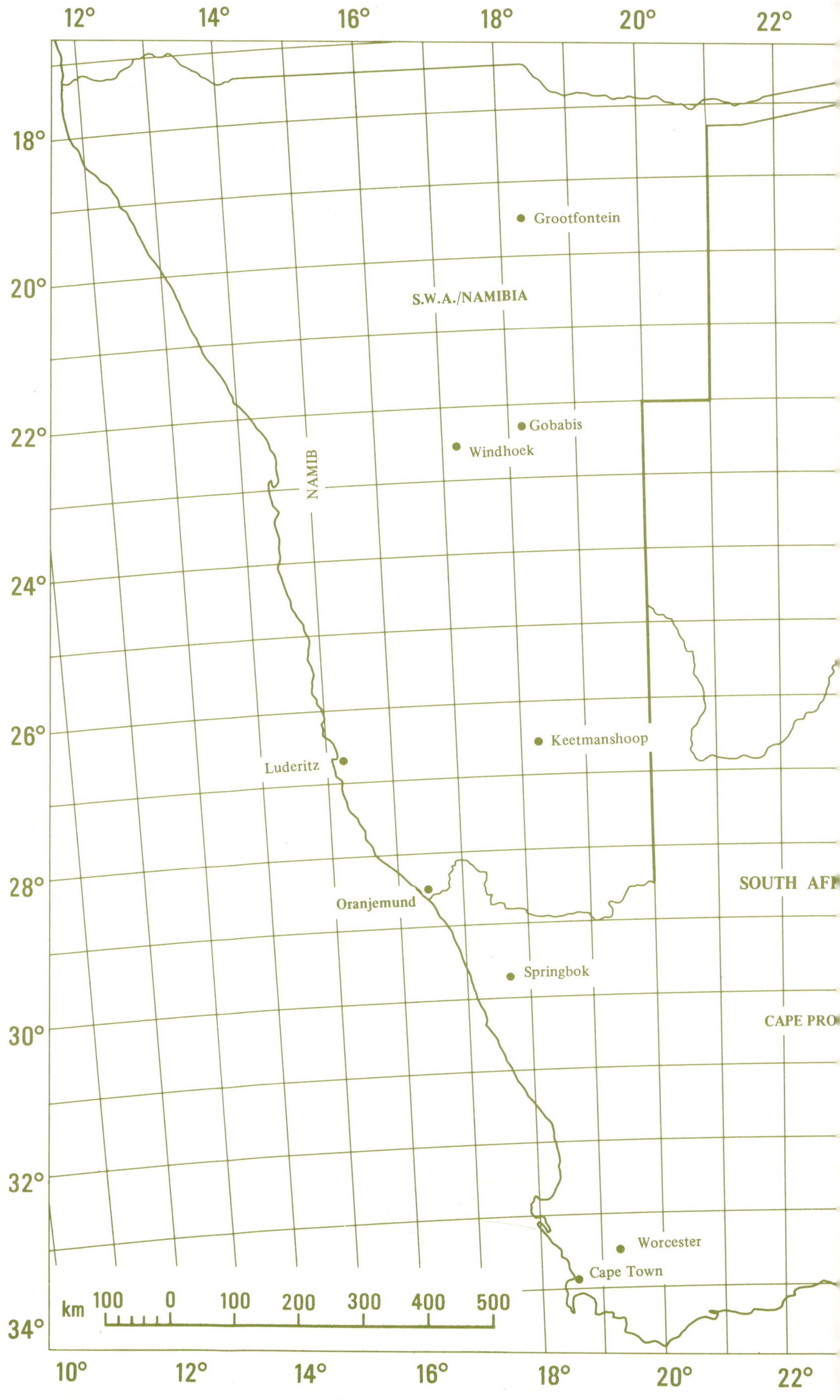
12°
14°
16°
18°
20°
22°
18°
20°
22°
24°
26°
28°
30°
32°
34°
Grootfontein
S.W.A./NAMIBIA
Gobabis
Windhoek
NAMIB
Keetmanshoop
Luderitz
Oranjemund
SOUTH AF
Springbok
CAPE PRO
Worcester
Cape Town
km 100 0 100 200 300 400 500
10°
12°
14°
16°
18°
20°
22°